USING GRAPHS AND TABLES

PETER H. SELBY
Director, Educational Technology
Man Factors, Inc.
San Diego, California

John Wiley & Sons, Inc.
New York · Chichester · Brisbane · Toronto

Editors: Judy Wilson and Irene Brownstone
Production Manager: Ken Burke
Editorial Supervisor: Winn Kalmon
Artist: Carl Brown

Library of Congress Cataloging in Publication Data:
Selby, Peter H.
  Using graphs and tables.

  (Wiley self-teaching guides)
  Includes index.
  1. Graphic methods—Programmed instruction. I. Title.
QA90.S46   001.4'22   78-25962
ISBN 0-471-05413-5

Printed in the United States of America
10 9 8 7 6 5 4 3 2 1

To the Reader

This book will show you how to use graphs and tables most effectively — in everyday life, on the job or in school. Working with graphs and tables can be fun — they can be interesting, and sometimes surprising, as well as useful. Briefly, graphs serve to:

- bring out the main numerical points about a subject;

- uncover facts that might be overlooked in text or in a table;

- summarize a cumbersome mass of data;

- add variety to text and tables, making material more interesting to look at and easier to read and understand.

Similarly, tables can serve to snap into perspective data that otherwise would be forever obscured in a fog of surrounding numbers.

Although we will look at ways of constructing some graphs and tables, we will focus on interpretation rather than construction. When you have finished this book, you should be able to recognize the major types of graphs, interpret them, and draw useful information from them, to apply for your own purposes.

The first chapter is an introduction to the world of graphs and tables. It should serve as a jumping-off point for your further exploration into the kinds of graphs that interest you most. If your interest is general, you should probably read all the chapters. However, if your time is short and your need specific, you may prefer to read just the first chapter and the chapter pertaining to your particular need.

The chapters in Using Graphs and Tables are divided into short, numbered steps called frames. Each frame presents some new information and asks a question or gives you a problem to solve, so you can learn to interpret graphs and apply the results most effectively. You should try to answer the question or solve the problem before you look at the given answer, which will follow the dashed line in each frame. (You may want to cover the book's answer as you read.) If your answer is not the same as that given, be sure you understand why before you go on, so you get the most out of later material. You will also find Self-Tests spotted throughout the book. These are your guideposts. If for any reason you cannot complete a Self-Test successfully, review the indicated parts of the chapter before you go ahead.

This book requires only a minimal mathematical background. Where a graph is based on a mathematical formula, the book explains how the formula is used in developing the graph and what useful aspects of the graph result from it. Interested readers are referred to other sources for more detailed discussion or deeper mathematical explanations.

Finally, don't try to hurry through the material. Set your own pace and go only as rapidly as your own aptitude and inclination permit. As always, the aim of teaching yourself should be learning, not speed.

La Jolla, California   Peter H. Selby
January 1979

Contents

ACKNOWLEDGMENTS

Graph on page 128(bottom) adapted from the George Washington University Newsletter.

Graphs and text on page 32(top) adapted from, and graphs on pages 32(bottom) and 40 reproduced from *Algebra Book One*, Revised Edition, by A. M. Welchons, W. R. Krickenberger, and H. R. Pearson. Copyright, 1956, by Ginn and Company (Xerox Corporation). Used with permission.

Graphs on pages 27, 52(middle), 124, 125(top), and 136(top) adapted with permission from *Graphical Analysis*, by Philip Stein. Copyright, 1964, Hayden Book Company, Inc., Rochelle Park, N.J.

Graphs on pages 39(bottom), 71, 72(top), 83(middle), 91, and 134 and table on page 69 from *Statistical Methods for the Behavioral Sciences*, by Allen L. Edwards. Copyright 1954, by Allen L. Edwards. Adapted and reprinted by permission. Holt, Rinehart, and Winston, Publishers.

Graph on page 129(top) adapted from the Institute of Life Insurance.

Graph on page 130 adapted from *Life Insurance Fact Book*, 1965

Graph on page 102(bottom) adapted from Los Angeles Chamber of Commerce Report.

Tables on pages 19 and 20(bottom) reprinted with permission of Macmillan Publishing Co., from *Statistics: A New Approach*, by W. A. Wallis and Harry V. Roberts. Copyright 1956 by The Free Press, a Corporation.

Graph on page 143 adapted from *Databook for Human Factor Engineers*, National Aeronautics and Space Administration, Ames Research Center, 1969.

Graphs on pages 82(bottom), 87(top), and 140(top) adapted from *Bioastronautics Data Book*, National Aeronautics and Space Administration, Government Printing Office, 1973.

Tables on page 20(top) and 51 reprinted with permission of the National Safety Council from *Accident Facts*, 1973 edition.

Graphs on pages 49(bottom) and 53, Copyright 1973 by the New York Times Company. Reprinted by permission.

Graph on page 5 reprinted from and graphs on pages 4 and 49 (top—original source *Steelways*) adapted from *How to Lie With Statistics*, by Darrell Huff, Pictures by Irving Geis. Copyright, 1954 by Darrell Huff and Irving Geis. By permission of W. W. Norton and Company, Inc.

Graph on page 83(top) adapted from San Diego County Chamber of Commerce Report.

Graph on page 126(bottom) adapted from Television Bureau of Advertising Inc., New York.

Graph on page 116(top) adapted from U. S. Bureau of Census.

Graphs on pages 101(bottom), 115, 126(top), and 127(bottom) adapted from U. S. Department of Agriculture.

Table on page 42 and graphs on pages 43-47, 52(bottom), 86, 87(bottom), 89, 90(bottom), 92(bottom), 93(middle), 94-96, 97(top), 102(top), 110(bottom), 111(top), 112(top), 113(top and bottom), 116(bottom), 117-122, 127(top), 128(top and middle), 129(bottom), adapted from *Standards of Statistical Presentation*, U.S. Department of the Army pamphlet 325-10, April 1966.

Graph on page 67(bottom) adapted from N.B.S., U.S. Department of Commerce.

Graph on page 133 adapted from U.S. Department of Health, Education, and Welfare.

Graph on page 112(bottom) adapted from U.S. Office of Management and Budget.

Graphs on pages 98, 99(top), 110(top), and 114 adapted and reprinted with permission of The Wall Street Journal, copyright 1974, Dow Jones and Co., Inc. All rights reserved.

Graphs on pages 136(bottom), 137, 138(top), and 142(middle and bottom) adapted from *Statistics* (a Wiley Self-Teaching Guide) by Donald J. Koosis, Copyright, 1972, by John Wiley and Sons, Inc. Reprinted by permission of John Wiley and Sons, Inc.

Graphs on pages 26(bottom) and 99 (bottom) adapted from *Psychological Statistics*, Fourth Edition, by Quinn McNemar Copyright 1969, by John Wiley and Sons, Inc.

CHAPTER ONE
Introduction to Graphs and Tables

With the growing use of graphs and tables to summarize data from every
branch of science, industry, business, and government, all of us need to be
familiar with their purpose and use. All of us? Yes, all of us. You need
not be a scientist, engineer, or business analyst to understand and use graphs
and tables. You may only want to be able to read the Dow–Jones industrial
average to get some idea of how the stock market is affecting your one share
of Amazing Electronics. Or to figure out, from the "pie chart" that comes
with your tax bill, where your tax money is going. Even closer to home is the
practical necessity of being able to read the table in your cookbook of roasting
times and temperatures for various weights and kinds of meat. Whatever your
need, the aim of this book is to help you learn to interpret tabular and graphic
information quickly and easily.

Of course, if you are involved in technological or business activity you
must be able to understand, use, and even make a variety of different kinds
of graphs and tables. Statistics, as applied to business and economics, relies
heavily on tables and graphs to summarize and simplify the presentation of op-
erational data. Statistical methods are also used extensively in psychology,
biology, genetics, medicine, and all branches of the behavioral sciences.
Statistics without graphs and tables would be like a sunset without the sun.
Likewise physics and chemistry—in fact, all the physical sciences—routinely
plot data on various types of graph grids to simplify information display.

Therefore, we are going to devote this chapter to considering some of
the kinds of graphs and tables you will encounter most frequently, their advantages and disadvantages, some important things to remember when preparing graphs, and some common misrepresentations of which you should be
aware. Specifically, by the end of this chapter you will be able to:

- recognize and describe some common fallacies and foibles of graphs;

- find the mean, median, and mode of a group of numbers;

- plot tabular data, read charts, and interpret the relationship between
  two variables;

- arrange a group of data in suitable tabular form;

- interpret trends in tabular data;

- identify the principal kinds of graphs;

- state the main advantages and disadvantages of graphs;

- distinguish between direct and inverse variation.

FALLACIES AND FOIBLES OF GRAPHS

1.   Disraeli's much quoted remark—"There are three kinds of lies: lies, damned lies, and statistics"—often appears to contain a great deal of truth. A lot of fun has been poked at statisticians, sometimes with good reason. This is not to damn all statistics or all statisticians. Far from it. We'd be boggled beyond belief if we were to try to sort out the meaning of the masses of data that presently are so succinctly summarized for us in handy tables and graphs. Nevertheless, caution is justified.

The most important thing to know about the interpretation of statistical data is that they <u>do</u> have to be interpreted. Seldom, if ever, do they speak for themselves. Statistical data in the raw simply furnish facts for someone to reason from. They can be extremely useful when carefully collected and critically interpreted. But unless handled with care, skill, and, above all, objectivity, statistical data may seem to prove things that are not necessarily true.

Many of us pass through two stages in our attitudes toward statistical conclusions. At first we tend to accept them and the interpretations placed on them, uncritically. But then we are misled so often by skillful talkers and writers who deceive us with correct facts that we come to distrust statistics entirely, feeling that "statistics can prove anything"—implying of course that statistics can prove nothing.

Somewhere in between complete skepticism and pure credulity there lies an attitude of informed receptivity that can enable us to use statistical tables and graphs to our own benefit without allowing them to use us.

Probably one of the most common misuses (intentional or otherwise) of a graph is the choice of the wrong scale—wrong, that is, from the standpoint of accurate representation of the facts. Even though not deliberate, selection of a scale that magnifies or reduces—even distorts—the appearance of a curve can mislead the viewer.

Consider the two graphs at the top of the next page, for example. Both curves represent the same relationship between the same two variables.

(a) Which would impress you most? _____

(b) Why? _____

_____

(c) What is the basic difference between the two graphs? _____

_____

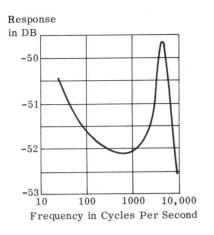

Frequency in Cycles Per Second

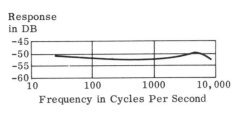

Frequency in Cycles Per Second

- - - - - - - - - - - - - - - - - - - -

(a) Probably the one on the left, regardless of whether or not you know what it signifies. The curve on the right looks so level as to be almost humdrum. Scale shrinkage (crowding the scale values closer together) could hide an important change in a graph that may go unnoticed. Similarly, scale expansion (spreading the scale values farther apart) could over-emphasize an unimportant change.

(b) Probably simply because of the wide area it covers—its extreme sweep and changes of direction.

(c) The graph on the left represents only a small band taken from the middle portion of the graph on the right and enlarged by stretching out the vertical scale. Thus what in the right-hand graph looked like a gentle curve with a slightly hooked right end becomes the steep and impressive curve of the left-hand graph.

2. You can see how easy it would be to give the reader whatever impression you wished to just by selecting a scale for your graph that resulted in its conveying the message you intended.

This can be accomplished nicely by dropping the zero or base line reference and replotting the curve on a greatly expanded scale. This is what has been done in the two graphs shown on the next page.

Can you tell what portion of the vertical scale in the chart on the right has been expanded to develop the chart on the left?

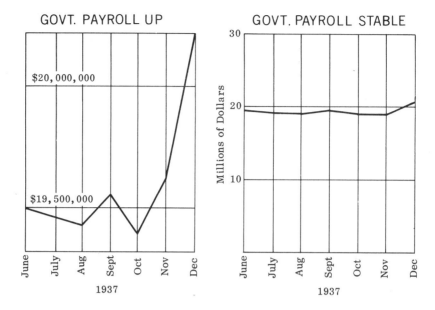

the vertical area of the chart that contains the curve, roughly the portion between $19 million and slightly over $20 million

This above example, which appears in Darrell Huff's excellent and amusing book How to Lie With Statistics, illustrates how easy it is to make an increase of less than four percent look like more than 400 percent.

3.   Another trick often used by those anxious to persuade others to their particular point of view has to do with averages.   The world—particularly in the realms of advertising and business—is very big on averages these days.   Yet the "average" reader has a very hazy notion of what the statistician means by the term.

What you and I generally think of when we use the word "average" is what the statistician calls the mean, that is, the arithmetical average of a group of numbers.   The number lying exactly in the middle he calls the median.   And the value (number) that occurs most frequently in the group he refers to as the mode.   The illustration at the top of the next page should help you get the idea.

Statisticians call these three kinds of average measures of central tendency, which is a way of saying that one value is typical of a distribution of measurements.

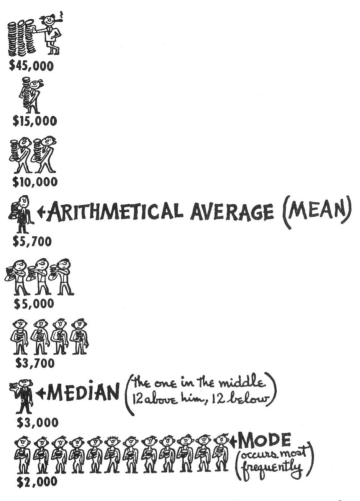

$45,000

$15,000

$10,000

◀ ARITHMETICAL AVERAGE (MEAN)
$5,700

$5,000

$3,700

◀ MEDIAN (the one in the middle, 12 above him, 12 below)
$3,000

◀ MODE (occurs most frequently)
$2,000

Because this is an important thing to know if you're going to be working with statistics—either as a consumer or producer of the same—let's look at another example.

We will assume you are the owner of a small business and that you have thirteen salaried employees. Let's assume further that there has been some grumbling in the ranks regarding salary levels. To counteract this sort of "subversive" talk you decide to compute the "average" salary and publish this information to your employees. Surely this will confirm your contention that you pay your employees well. But which average?

The figures at the right represent the salaries you pay each of your employees. Deciding it would be best to calculate all three types you arrive at the following.

Salaries Paid
$13,500
13,500
13,500
9,000
7,000
6,500
5,000
4,500
4,500
3,500
3,500
3,500
3,500
$91,000

(a)   The <u>mean</u> (total salaries divided by the total <u>number</u> of salaries, that is, $91,000 divided by 13) turns out to be _____.

(b)   The <u>median</u> (salary in the middle) is

_____.

(c)   And the <u>mode</u> (salary figure that appears the most times) proves to be _____.

- - - - - - - - - - - - - - - - - - - -

(a) $7,000; (b) $5,000; (c) $3,500

4.   Now which of these averages are you going to use to convince your employees of your liberality as an employer? The arithmetic average is $7,000. However, the middle salary is $5,000, which is $2,000 less. And, as a matter of fact, the salary you pay to the greatest number of your employees is $3,500.

Which would <u>you</u> choose to accomplish your intent? _____

- - - - - - - - - - - - - - - - - - - -

I don't know about you but I would choose the highest one, namely, the mean—$7,000.

5.   Now, however, suppose one day you are approached for a donation by a representative of your local united charity drive. This fellow has heard of your liberal policy in paying good salaries (probably from one of your employees—<u>not</u> one of the ones who read your report on the mean, but one of those earning $13,500) and wants you to make a company contribution comparable with those high salaries. (<u>He</u> has the $13,500 type salary in mind, of course.)

Which average salary are you going to quote him in defense of the modest contribution you plan to make? _____

- - - - - - - - - - - - - - - - - - - -

the mode—$3,500

6.   The interesting thing about the two different figures you used to support the position you wanted to take at the moment is that both—in fact, all three— are absolutely valid, and you are quite justified in using them. The fallacy— if there is one—is that the person to whom you are quoting your statistical average rarely knows enough to ask what <u>kind</u> of average you are using! So beware that <u>you</u> ask the right questions.

Statistics for the most part represent an analytical treatment of samples. And for the sampling procedure to have any validity it must be unbiased. Yet we are constantly being asked (most frequently in advertising) to accept as valid statistics based on samples with a built-in bias. Opinion polls, for example, represent a running battle against bias, but despite the fact that they are usually run by well qualified and reputable polling organizations, you should bear in mind that this battle can never be "won."

Let's consider a simple and fairly common case. You are a graduate (20 or 25 years ago) of a major university. One day you receive a questionnaire from the alumni secretary asking you to supply certain personal information about yourself, including your present salary. You decide to comply with the request because you are promised a copy of the final summary and analysis if you do, and you are curious to know how your situation compares with that of your former classmates.

Two months later you receive the report. One figure catches your eye, namely, the <u>average</u> (the mean, in this case) salary, which turns out to be $26,218—at least $5,000 more than you are making! How come? you ask yourself. I thought I was doing OK. Am I such a financial flop compared to the rest of the class?

Then you get suspicious of this reported average. Just how reliable is it? So, curiosity mingled with self-justification, you do some analytical thinking about how this result was arrived at, and you write down some of your thoughts.

Can you think of some of the things <u>you</u> might have written down had this

really happened to you? (Maybe it has.) _____

_____

_____

_____

- -

Here are a few ideas that may have occurred to you. (1) Why such an exact figure for average salary? It is rather unlikely that the average income of any far-flung group is going to be known to the dollar. There is an air of suspicious specificity about this number. (2) It would have been impossible for the questionnaire to reach <u>all</u> (living) members of the class since the addresses of a great many of them must be unknown after a 20- to 25-year interval. (3) The average salary figure shown could only be based on what respondents <u>said</u> they earned. Doubtless many lied. (4) Nothing was said about the percentage of return of completed questionnaires. Many people probably didn't bother to complete and return the form sent them. (5) Individuals in low income brackets are likely to be most reluctant to report their salaries. On the other hand the more affluent types would have both the time and the inclination to let others know how well they have succeeded in their climb up the financial ladder. Therefore the return would reflect a bias toward the more affluent members of the class.

7.   Note that the survey sample in frame 6 simply represents living members of the class whose addresses are known and who are willing to say how much they earn.   Therefore it can hardly be considered an unbiased sample.   It also illustrates the fact that a purely random sample (the only kind that can be examined with complete confidence) is very difficult to obtain.   So you must always remember that any statistical sample has a certain amount of distortion in it.   The question is how much?   You can answer that best with some common sense skepticism and with close analysis of the way the sample was obtained.   This is, of course, harder to do in fields you're not familiar with.

Following closely upon the heels of the biased sample is the inadequate sample.   A prime example of this, and of which every TV viewer is (painfully) aware, is the "fewer cavities" motif.   Any group that keeps track of its cavities for six months and then switches to some name brand toothpaste is going to notice one of three things: more cavities, fewer cavities, or about the same number of cavities.   If this test is repeated often enough, sooner or later, by chance, some group will show a significant reduction in the number of tooth cavities.   That's the result that will launch a major advertising campaign!

A similar game can be played with pictographs, also.   (Pictographs are pictorial displays of statistical information.)   Below are two examples.

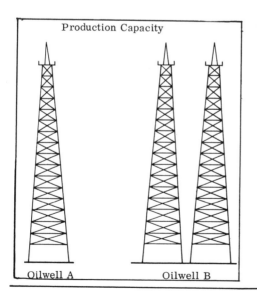

Usually, if one small picture of an oil derrick is used to represent the production capacity of oilwell A, we would use two such pictures to represent the capacity of an oilwell that has twice the capacity of A, namely, oilwell B. Such a comparison is shown in the example at the left on page 8.

But suppose we decide to portray the production capacity of oilwell B by using a picture of an oil derrick that is twice as tall as the one representing oilwell A, as shown in the example at the right. Nothing wrong with this, is there? The scale is the same: twice the height represents twice the capacity.

Do you see any difference between the two methods in conveying the correct impression to the reader? _____

_____

_____

- -

Although the single derrick representing oilwell B is (correctly) twice as <u>tall</u> as the one representing oilwell A, it also is (incorrectly) twice as <u>wide</u>. It therefore occupies not twice but <u>four times</u> as much area on the page. The numbers still say <u>two</u> to one, but the visual <u>impression</u>, which is the dominating one most of the time, says the ratio is <u>four</u> to one. Watch out for this trick! The visual inference can easily take precedence over the stated numbers.

There are numerous other ways in which the unscrupulous or the ignorant can use statistics to misrepresent facts. Sometimes if they can't prove what they want to prove, they may demonstrate something else and pretend (or think) it's the same thing. Often just drawing a graph of two variables implies a correlation where none exists. Since few of these appear in formal table or graph form we won't go into them here. However, that's why we need to know how to interpret them properly—what they mean and <u>don't</u> mean, their possibilities and limitations.

Having taken a quick look at a few of the common <u>fallacies</u> of chart-making, let's turn our attention to some of the correct procedures and legitimate uses of tables and graphs.

HOW GRAPHS AND TABLES CAME TO BE

8. As far as we know, William Playfair, an English statistician, was the first person to apply charts and graphs to finance. In 1786 he wrote, "I have succeeded in proposing a new and useful mode of stating accounts . . . . As much information may be obtained in five minutes as would require whole days to imprint on the memory . . . by a table of figures."

If William Playfair was concerned about reducing masses of data to compact, readily understandable form, think how much greater are those masses of information today!

We commonly associate graphs and tables with the field of statistics. And although contemporary statisticians would probably give the term a broader meaning, we can think of statistics as numerical descriptions of the quantitative aspects of things as determined by counts or measurements. For example, statistics on the makeup of the students in a class might include a count of the number of students, and separate counts of the numbers of members of various kinds—number of males and females, number over and under 20 years, and so on.

Suppose you wanted to find out for yourself what the temperature changes look like throughout the year in the area where you live. Having mounted a thermometer outside some convenient window (where you could see it without having to go outside in the cold), and promised yourself you would faithfully make regular observations, you note the outside temperature each day, at the same time, for a whole year. At the end of the year you are the proud possessor of 365 temperature readings. Now, what are you going to do with them? You could type them up neatly, frame them, and hang them on the living room wall, but they probably would not make very interesting reading. Also, although you could perceive a general trend toward higher temperatures in summer and lower in winter (not exactly unexpected), there are so many numbers and they look so much alike that you may not feel you know any more than you did before you took all those readings.

So, what to do? First, you should decide what it is you really want to know about your local weather. What will this accumulation of temperature readings tell you that you didn't know before? Before you started you already knew it was colder in winter and warmer in summer, so that isn't news. Let's suppose you want to know the average monthly temperature throughout the year. And you would like this information to be based on your actual temperature readings—that is, specific numerical values—not on someone else's guess or your own vague impressions. How do you get this?

Do you remember how to average a group of numbers? If so, describe briefly below how you would go about it. If not, review the answer following the dashed line. _____

_____

- -

Your answer should say something like this: Add all the temperatures for each day of the month together and then divide the result by the number of days in the month. (For example, the average of 3, 10, 6, and 5 is 6. The sum of these numbers, 24, is divided by the quantity of numbers in the list, 4, to obtain their average, 6.)

9.   This is the correct procedure, all right. As we learned in frame 3, a statistician would refer to this as finding the mean of the group of numbers, but we will stick to the more familiar term "average" for now.

Let's suppose you average the temperatures for each month of the year and arrive at a series of twelve numbers. Now, what will you do with these numbers?

A logical thing to do would be to arrange them in chronological sequence. Suppose we do so (I will supply the numbers). This gives us the list of months and temperatures at the right. What we have created is a <u>table</u>, which is simply a systematic arrangement of data, usually in rows and columns, for ready reference. Notice that we have drawn lines around our data to box it in. This is not necessary, strictly speaking, but it usually looks better and makes it somewhat easier to read related values, especially if the table contains consider- able data.

Month	Temperature (°F)
January	40
February	40
March	42
April	47
May	53
June	57
July	62
August	62
September	59
October	52
November	44
December	41

Below is a graph of the data in our table above. Study this graph for a few moments and then see if you can answer a few questions about it.

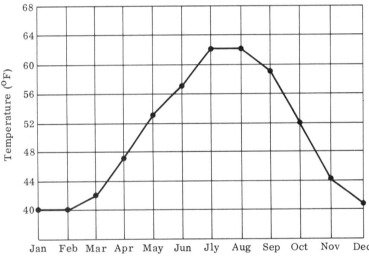

(a)   What are the values on the vertical (left) scale?   _____

(b)   What are the values on the horizontal scale?   _____

(c)   What is one unit of the vertical scale equal to?   _____

(d)   How many scale units (squares) apart are the months?   _____

- -

(a)   temperature in degrees Fahrenheit (°F)
(b)   the months of the year
(c)   four degrees
(d)   one

10. In order to be able to plot the values in our table, and thus transform it from a table into a graph, we used what is often termed a grid system. A grid system is simply a network of straight lines that are at right angles to one another, much like the streets in most cities or the wires in a cookie-cooling rack or wires in a screen.

You will notice that the grid-lines in our graph (or chart, as it also is called) are evenly spaced, both vertically and horizontally. Usually the spacing between the vertical lines is uniform and the spacing between the horizontal lines is uniform. But the spacing between the vertical and horizontal lines need not always be the same, as we shall see later.

Because the vertical and horizontal lines are at right angles to each other this kind of grid system is known commonly (in mathematics) as a rectangular or Cartesian (after René Decartes, the man who developed it) coordinate system.

Let's review what we know so far about the grid system.

(a)   The chief characteristic of a rectangular coordinate system is that the

vertical and horizontal grid-lines are at _____ to one another.

(b)   The spacing between the vertical and horizontal grid-lines usually is

_____ .

(c)   The spacing between the vertical grid-lines need not be the same as that

between the horizontal grid-lines. (True or False) _____

- -

(a) right angles; (b) uniform; (c) true

11. Below is our graph again. Let's see how we converted the data in our table of temperatures into the curve we see on the graph.

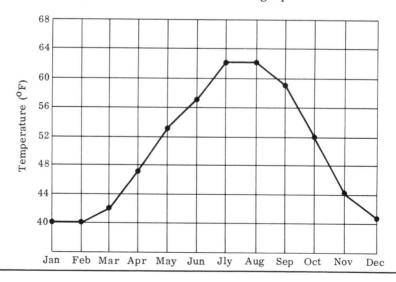

Since the vertical scale represents temperature values in degrees Fahrenheit (the temperature scale in which we took our readings from the thermometer), it is apparent that the numbers we find there indicate increasing temperatures. (Notice that the temperature scale on our graph does not start at 0°, as it would be customary to do; instead it was started near the lowest reading in order to reduce the vertical size of the graph.) Similarly, the vertical marks above each month represent successive positions along the time scale. Therefore, in order to establish a one-to-one relationship between any particular month and its average temperature value we simply select a month, note its temperature figure, find the name of that month at the bottom of the chart, and mark a point on the vertical line centered above the month opposite the correct temperature value.

Try it with March. Our table tells us that the average temperature for the month was 42°. Following the line above "Mar" to the point opposite 42° (midway between 40° and 44°) on the vertical scale we find a dot. This dot represents the meeting point for the two quantities, time and temperature. Similarly, the point on the center vertical line above "Jly" shows that the average temperature during July was 62°, and so on. One thing is certain from both the table and the graph: you live in a cold climate!

Connecting the twelve dots gives us the line graph we see. Can you think

of any advantages this graph has over the table? _____

_____

_____

_ _ _ _ _ _ _ _ _ _ _ _ _ _ _ _ _ _ _ _

Some advantages you may have observed are: (1) The curve gives you a much quicker impression of how the temperature varies throughout the year. (2) The temperature scale allows you to read intermediate temperature values between months. And even though these do not represent specific observations, they do furnish an approximation of the mid-temperature between monthly averages. (3) The curve shows clearly the highest and lowest temperatures—their values and when they occurred.

Graphs have other useful features that will be more evident in later examples. We will discuss some of these features in the sections that follow.

DATA RELATIONSHIPS

12. The main purpose of tables and graphs is to help us understand the relationships between and among various kinds of data. (Incidentally, data is a plural term; datum is the singular.) "Raw data" (data derived directly from observation) seldom are very meaningful. Such data need to be analyzed, simplified, and presented in a way that is easy for the reader to understand.

For example, the raw data for our temperature graph consisted of 365 daily readings. Although we didn't actually look at these readings, you can readily appreciate that by themselves, the readings would do little to illuminate the month-to-month pattern of temperature changes. In fact the daily variations in temperature would tend to obscure any monthly or seasonal patterns or trends.

Not until we computed the mean averages of temperature for the months and put these in tabular form (frame 9) did the pattern emerge. We then were able to identify not only the specific temperature means for each month but also the rising and falling trend throughout the year. Using the table to plot a graph of the monthly mean temperature values furnished a picture that reinforced our understanding and allowed us to see at a glance the high and low temperature points for the year. Thus, graphs help us see relationships that might otherwise pass unnoticed.

Whether you are presenting data yourself or interpreting someone else's, you should remember that the mode of presentation selected communicates the results of a whole chain of statistical operations, including collection, compilation, processing, and analysis of data. What appears on a few pieces of paper often is the only useful product of many hours spent assembling and summarizing the basic data.

Basically, the statistical analyst has three forms of presentation available to him: text, tables, or graphs. Text (which we really are not concerned with in this book) is not suitable for most statistical reports because only a relatively small number of figures can be included without confusing the reader. Although tables are the backbone of statistical reporting, giving the data which is further summarized in graphs, all-tabular reports seldom are readable. All tables and no charts or text would make a very dull report. Because graphs are attractive and forceful one is tempted to rely on them altogether in showing data relationships. However, all-graphic presentation is unsatisfactory because it lacks a smooth flow of thought and also because it leaves unanswered too many questions that can easily be answered by text or tables. So all three modes of presentation have their place and are best used in combination.

Even before data are collected and compiled it is a good idea to decide what purpose they are intended to serve, what audience they are to reach, and in what form they ultimately will be presented. This in turn will lead to decisions—certainly more intelligent ones—as to what data to include, which to emphasize most strongly, which to present in detail and which in summary form, which to present in graphs and which in tables.

Suppose you are given an assignment to find out something about the willingness of airline passengers to fly again after a flight in which certain comfort factors were assessed. You interview each of the deplaning passengers and with the aid of a questionnaire obtain two kinds of information: their degree of comfort and their willingness to fly again. Your hypothesis is that there is some kind of correlation between the comfort they experienced on the flight just completed and their willingness to repeat the experience.

In order to be able to compare levels of comfort with levels of willingness to fly again, your questionnaire provided five degrees of comfort and five of willingness. So you now have a collection of data interrelating two variables (comfort and willingness to fly) assessed at five levels. The first logical step in sorting out, relating, and analyzing these data would seem to be to put them in tabular form.

The five categories of comfort are: very comfortable, comfortable, neutral, uncomfortable, and very uncomfortable. The five categories of willingness to fly are: eager, no doubt, some doubt, prefer not, and never. See if you can prepare a tabular form on which to enter your data, using the categories just named as the headings for your rows and columns. (If you haven't done anything of this kind before, don't be alarmed; just do your best.) There is space below for your table categories.

- - - - - - - - - - - - - - - - - - - -

Distributions of Willingness to Fly Again by Comfort Level.

	Willingness to fly				
	Eager	No doubt	Some doubt	Prefer not	Never
Very comfortable					
Comfortable					
Neutral					
Uncomfortable					
Very Uncomfortable					

Your (blank) table should look something like the one shown above. You could, of course, have let the "willingness to fly" categories serve as row headings rather than column headings, and the comfort categories as row headings. This wouldn't be incorrect. However, the outcome factor customarily is shown as the column headings. Since we are assuming willingness to be an outcome of comfort, we have assigned willingness to fly as the column heading(s). Notice, in either case, that the <u>levels</u> of response for each factor (variable) are arranged in descending order of intensity, that is, from <u>most</u> comfortable to <u>least</u> comfortable (actually, very uncomfortable), and from <u>most</u> willing (eager) to <u>least</u> willing (never). This arrangement helps us to observe most easily any relationship that may exist among the data. (You would not be expected to title your graph; in actual use, however, every table or graph should have a title or explanatory caption.)

13. Now it is time to fill in some numbers. We could use the actual numbers (quantity) of individuals reporting in each category, but these usually are not

as meaningful as percentages. Therefore, considering the first category of
comfort (very comfortable), we divide the total number of those reporting
their previous flight as "very comfortable" <u>into</u> the number of those who also
categorized themselves as <u>eager</u> to fly again. The quotient turns out to be
.67, which means that 67% of those who were very comfortable during their
last flight are eager to fly again. Similarly, dividing the number of those who
"no doubt" would fly again by the total of those who were very comfortable,
we find that 33% wind up in this category. Following this same procedure for
each of the other rows (i.e., categories of comfort) gives us the values shown.
(If you have trouble with percentages, just imagine that 100 passengers reported in each comfort category; 67 of the 100 who were very comfortable were
eager to fly again, 33 of them "no doubt" would fly again, and so on.) The results of this dividing (which we haven't burdened you with) are shown in the
completed table below.

Distributions of Willingness to Fly Again by Comfort Level. *

	Willingness to fly				
	Eager	No doubt	Some doubt	Prefer not	Never
Very comfortable	67	33	0	0	0
Comfortable	20	74	5	1	0
Neutral	8	73	12	6	1
Uncomfortable	3	49	29	17	2
Very uncomfortable	3	13	24	37	23

*Table entries are percent of row total.

What, if any, trend or relationship between the data do you observe?

_____

_____

- -

The table clearly shows a relationship between the comfort rating for the trip
just completed and the person's willingness to fly again. Later we'll discuss
more fully what to look for (or how to spot trends) in analyzing a table.

14. Are we through with these data? Have we obtained the clearest picture
possible of the relationships that exist between them? Not quite. Although
we can observe, roughly, a relationship from the table, we won't really com-
prehend the <u>nature</u> of the relationship until we graph it.
     Since we are primarily interested in knowing the percent of passengers
with no doubts about flying again, we will pool the "eager to fly" and "no doubt"
categories and plot the resulting percentages against the comfort ratings
(shown as ranging from 1 to 5 for simplicity) and obtain the plot shown at the
top of the next page. This can be viewed as a percent-satisfied curve. Using
it we can predict the percent of passengers willing to fly again from their comfort ratings.

What do you think this chart is saying to the airlines?

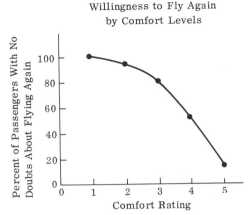

Willingness to Fly Again
by Comfort Levels

_____

_____

_____

_____

- - - - - - - - - - - - - - - - - -

The obvious message to the airlines is, if you want to have over 90% of your passengers with no doubts about flying again, provide a flight that yields a comfort rating of two or better.

DATA RELATIONSHIPS IN TABULAR FORM

15. In this section we are going to take a closer look at the characteristics and usefulness of tables as a means of presenting data relationships.

As you already are aware, the presentation of data in tabular form is a basic statistical device. A statistical table is a systematic arrangement of numerical data in columns and rows. Its purpose is to show quantitative facts clearly, concisely, and effectively. It should facilitate an understanding of the logical relationships among the numbers presented.

Tables are used in the compilation of raw data, in the summarizing and analytic processes, and in the presentation of statistics in final form.

A good table is the product of careful thinking and hard work. It is not just a package of figures put into neat compartments and ruled to make it look more attractive. It contains carefully selected data put together with thought and ingenuity to serve a specific purpose.

Tables are, as mentioned before, the backbone of most statistical reports. They provide the basic substance and foundation on which conclusions can be based. They are considered valuable for the following reasons:

Clarity—they present many items of data in an orderly and organized way.

Comprehension—they make it possible to compare many figures quickly.

Explicitness—they provide actual numbers which document data presented in accompanying text and charts.

Economy—they save space, and words.

Convenience—they offer easy and rapid access to desired items of information.

There are, of course, disadvantages of tables. Some of these include the following:

> Uninviting—tables often look like difficult reading matter so many people ignore them.

> Undramatic—significant relationships often are hard to find and comprehend.

> Too specific—tabular data, because of their formal appearance, tend to make crude estimates and projections seem more precise than they actually are.

Much of the art of tabular presentation lies in the ability to maximize the advantages and minimize the disadvantages. Although the above points are best borne in mind when creating tables, they also are important to know about when interpreting tabular information.

Arranging row and column headings in order of increasing (or diminishing) sensitivity, where the data permit this, often helps to reveal trends that otherwise might not be noticed. Let's return to our "willingness to fly" table again (below).

Distributions of Willingness to Fly Again by Comfort Level. *

	Willingness to fly				
	Eager	No doubt	Some doubt	Prefer not	Never
Very comfortable	67	33	0	0	0
Comfortable	20	74	5	1	0
Neutral	8	73	12	6	1
Uncomfortable	3	49	29	17	2
Very uncomfortable	3	13	24	37	23

*Table entries are percent of row total.

What trends do you observe in the "prefer not" column and the "very uncomfortable" row? Are these in accordance with expectation?

_____

_____

- - - - - - - - - - - - - - - - - - - -

The percentage of those who "prefer not" to fly again increases steadily (directly) with diminishing comfort—or, if you will, increase in discomfort—trom 0% to 37%. Among those who were very uncomfortable, the percentage of those reluctant to fly increases steadily from 3% who are "eager" to 37% who "prefer not." Both trends certainly are in accord with our expectation since the more uncomfortable one has been on his last flight, the less likely he would want to fly again. Similarly, among those who were "very uncomfortable," we would expect very few who would be eager to fly again and an increasing number who would be reluctant to fly again.

16. In order to spot trends in tabular data, look for fairly consistent increases or decreases in the columns or rows. Then consider what those trends imply about the subject matter of the table. Here is another example.

Income from Legal Practice, by Law School Attended,
Chicago Lawyers, 1947

Law School Attended	Mean Income	No. of Cases	Average Experience
University of Michigan	$18,523	22	21.1
Harvard University	18,294	46	18.0
University of Chicago	11,306	116	18.1
Northwestern University	11,247	88	16.2
Chicago-Kent	10,130	129	20.1
Chicago College of Law	9,512	20	28.3

The table above contains data relating to the income of a group of Chicago lawyers based on the law school attended. If you intended to become a lawyer and wished to choose a school on the basis of future income prospects, would the results reported here convince you that you should go to one of the two top schools in this table? Why? _____

_____

- - - - - - - - - - - - - - - - - - - -

Not necessarily. The data certainly wouldn't help you choose between Harvard and Michigan. But even more important, the data don't really show whether the law school attended bears any relation to earning capacity. You would want to find out whether any hidden factors needed to be allowed for. (Remember the alumni questionnaire we discussed in frame 6 and some of the built-in fallacies?) Earning capacity certainly includes connections of all kinds, and it might be that students with "good connections" are more likely to attend certain schools. Also students from Chicago who go to Harvard, or any other expensive or distant college, are probably already better off financially.

17. Glance back to the table shown in frame 16 and you will notice that it consists of one column of row headings and three columns of numerical data. The table at the top of the next page differs in that the column headings are repeated in the right half of the table. This arrangement has no particular significance other than the fact that it probably suited better the layout of the page on which it originally appeared. The material in the right half could as easily have been continued under the left half, assuming there was sufficient vertical space on the page. This is a matter which table makers must take into account when preparing materials for publication.

But since we are more interested in interpreting the content of the table than analyzing its format, study the data at the top of the next page for a few moments and then see if you can answer the questions which appear below the table.

Accidental Death Rates by Nation

Nation	Year	Deaths	Rate*	Nation	Year	Deaths	Rate*
Dominican Rep...	1970	747	18.4	Italy.............	1970	24,933	46.5
Paraguay........	1969	442	19.1	Nicaragua........	1969	913	47.6
Philippines.......	1969	7,992	21.5	Netherlands......	1971	6,235	47.9
Thailand........	1970	9,409	27.7	Poland...........	1971	15,711	48.0
Salvador.........	1970	1,158	32.8	Denmark........	1970	2,375	48.2
Spain............	1970	11,879	34.8	Norway..........	1970	1,975	50.9
England, Wales...	1971	17,158	35.3	Canada..........	1970	11,378	53.2
Costa Rica.......	1969	596	35.4	Australia.........	1971	6,802	53.3
Guatemala.......	1969	1,806	36.0	Portugal.........	1971	4,817	54.3
Yugoslavia.......	1970	7,477	37.3	New Zealand.....	1971	1,569	55.0
Israel...........	1971	999†	38.5	United States.....	1971	115,000	55.8
Greece..........	1970	3,460	39.4	Hungary.........	1971	5,935	57.3
Panama.........	1970	583‡	39.8	Czechoslovakia...	1970	8,466	58.5
N. Ireland.......	1971	610	39.9	East Germany.....	1970	10,015	58.7
Japan...........	1971	41,936	40.1	Finland..........	1970	2,745	59.5
Ireland..........	1971	1,239	41.7	Iceland..........	1970	127	60.5
Scotland.........	1971	2,215	42.4	Switzerland......	1970	3,785	61.2
Sweden..........	1970	3,420	42.5	West Germany....	1970	38,997	63.3
Uruguay.........	1970	1,228	42.5	Belgium.........	1969	6,366	66.0
Bulgaria.........	1971	3,657	42.8	South Africa.....	1969	2,688**	72.1
Taiwan..........	1970	6,545	45.1	France..........	1970	37,723	74.3
Venezuela........	1971	4,966‡	46.2	Austria..........	1971	6,169	82.7

(a) Which nation shows the largest number of accidental deaths for the year 1970? _____ How many? _____

(b) Which nation shows the largest number of accidental deaths for any year? _____ Which year? _____

(c) Which nation shows the highest accidental death rate in any of the three years reported? _____ Which year? _____

– –

(a) West Germany, 38,997; (b) United States, 1971; (c) Austria, 1971
(Did you notice that the countries were listed in increasing order by death rate?)

18. Returning again to the matter of lawyers, their education and income (apparently a very popular subject for statistical analysis!), the table below presents data concerning income from legal practice according to education, but from a slightly different point of view than we considered in frame 16.

Year of Admission to the Bar	High School, No College		College Graduates	
	Mean income	No. of cases	Mean income	No. of cases
Prior to 1910	$13,559	17	$22,132	19
1910–1914	19,188	16	21,705	22
1915–1919	10,577	26	19,053	19
1920–1924	17,100	10	16,095	21
1925–1929	8,206	34	13,066	76
1930–1934	5,500	1	12,111	81
1935–1939	——	0	9,050	107
1940 and after	2,000	1	4,696	97

If you wanted to be a lawyer and were trying to decide whether to get your training in college or by an office apprenticeship, with income-producing ability as the sole criterion, would you find the results of this table convincing for

an office apprenticeship? Why? _____

_____

- -

Probably not, because the table shows that in every experience category except one (1920-1924) the college graduates have a higher income than lawyers admitted to the bar with no college education.

DATA RELATIONSHIPS IN GRAPHIC FORM

19. When only a few key numbers are to be presented it often is desirable to show them only in the text. This is especially true if the figures lack comparability and need explanation. And as we have seen, tables represent the main mode of presenting statistical data in numerical form. However, graphs often are most effective for showing simple relationships, trends, and general magnitudes.

Now let's discuss some of the basic types of graph you are apt to encounter and therefore need to be able to interpret. Since the remaining chapters of this book will deal with each type of graph in greater detail, we will simply introduce the broad categories to you. (Note: See the Appendix for a detailed chart showing the graphs we discuss in this book.)

There are many types of graph and many ways of classifying them, but for our purposes we will divide them into four general categories:

    (1) Line graphs;
    (2) Surface graphs;
    (3) Bar graphs;
    (4) Special graphs (i.e., combinations or variations of the above plus some unique types).

Let's consider <u>line graphs</u> first. The graph we used to show the relationship between average temperatures and the months of the year in which they occurred was a line graph. But it was a particular <u>type</u> of line graph. Line graphs may be subdivided into five categories, defined below and on the next page.

Straight-Line: Consists of one or more straight lines representing a linear relationship between two quantities.

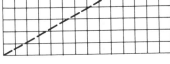

Curvilinear: Consists of a curved line representing a nonlinear relationship between two quantities.

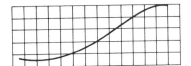

Zigzag (connected line segment): A curve made by drawing a line directly from each plotted point to the next. Shows whether changes from point to point are gradual or abrupt.

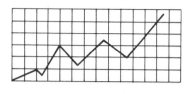

Step: A graph made by drawing a horizontal line through each point and connecting the ends of these lines by vertical lines. Often used to show averages or other measures that apply over periods of time.

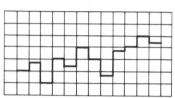

The fifth group of line graphs are known as Special Scale graphs. These differ sufficiently from one another to make it impractical to show you a typical example here; you will find these covered in the next chapter.

From the information above and on the previous page, into what general category would you say our temperature chart falls? _____

- -

Zizag. Although the points on our chart marking the average temperature for each month didn't result in quite the sawtooth curve we see in the example, it was made by drawing straight lines from each plotted point to the next.

20. The second category of graphs are known as surface graphs. Surface graphs differ from line graphs in that, although the curve (shape of the graph) may be straight-line, curvilinear, zigzag, or step, we are interested primarily in the area under the curve. You will recognize surface graphs by the fact that they are shaded under the curve. Let's take a look at a few of them so that you'll get the idea.

Simple zigzag: This is simply a single-curve chart with the space between the curve and the base line shaded to form a surface.

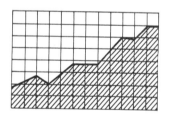

Simple step: This is essentially the same as the simple step line curve but with the area between the curve and the base line shaded to form a surface.

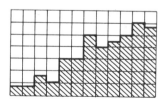

Subdivided zigzag: Also known as a layer or strata chart, this graph often is used to show how the component parts of a time series combine to make the total.

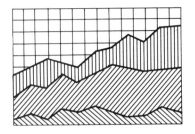

To which of the other charts would you say this last graph is closely related? _____

- -

The simple zigzag-surface graph. The subdivided zigzag-surface graph is simply a combination of several simple zigzag-surface graphs.

21. Before we leave the surface graphs let's take a look at one more variety, namely, the curvilinear-surface graph.

Curvilinear: This is basically a non-linear, simple curve with the area between the curve and the base line shaded to form a surface. Notice that (in this case) it has definite limits on the horizontal scale.

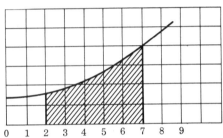

See if you can write the upper and lower limits here. _____

- -

2 and 7. These numbers mark the beginning and end of the shaded area and therefore constitute the limits.

22. The third type of graphs are known as bar graphs. You probably have seen more of these than any other kind, since they are used in all types of financial reports by corporations, civic organizations, and governmental departments, as well as for many other purposes.

There are two basic types of bar graphs: vertical (or columnar) and horizontal. They are shown on the next page.

<u>Vertical</u> (simple column): Consists of a series of vertical bars, each extending from the base line to a particular vertical height.

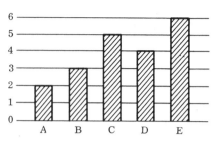

<u>Horizontal</u> (simple bar): This chart is merely a series of horizontal bars drawn to the right of a common base line.

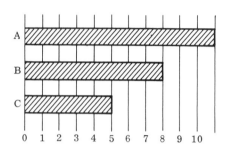

There are, of course, many variations of these two types of bar graphs—grouped column, subdivided column, deviation column, bar-and-symbol, subdivided bar, and so forth. We will get into these in Chapter 4. It is enough now that you have some idea of what bar graphs look like (if you didn't already) and how they are plotted.

The other name for a vertical bar graph is _____.

- -

columnar; (This is important because many people <u>always</u> refer to a vertical bar graph as a columnar graph. There is nothing wrong with this, of course. It's purely a matter of preference in terminology. However, by attaching the name <u>bar</u> we identify this type of graph with its family category.)

23. The fourth general category of graphs we have simply called <u>special graphs</u>. These are subdivided into two general categories: <u>combination graphs</u> and <u>other graphs</u>.

Combination graphs may consist of any of a great many combinations of line, surface, or bar graphs, such as zigzag and step, surface and curve, or column and step. Here and on the next page are two examples.

<u>Zigzag and step curve chart</u>: Combines zigzag curves and step curves for a more effective chart.

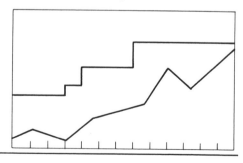

Surface and curve chart: This combination can be used in many ways.
Thus, a step curve representing a
standard or requirement can be combined with a simple surface chart
showing actual performance.

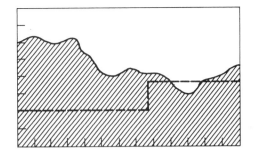

Have you any idea what a column
and step curve chart would look like?
See if you can draw one in the space
provided at the right. All you have
to do is combine some form of columnar (vertical bar) chart with some
form of step curve in the same graph.

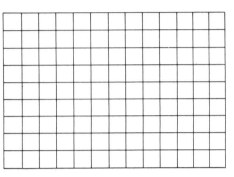

- - - - - - - - - - - - - - - - - - - -

Your graph might look something like
this. As long as you have included
some type of step curve with a vertical bar graph you're on the right track.

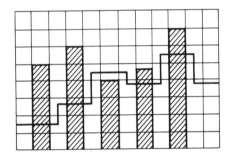

24. As you would suspect, "other graphs" is a catch-all for a group of graphs
that don't really lend themselves to classification under any of the previous
headings.

Among these are three that you surely will recognize: the pictograph,
the histogram (although you may not know it by that name), and the circle
graph, or pie chart. These are shown on the next page.

Pictograph: The pictograph is simply a graphic representation of a statistic using small, simple pictures to represent a certain number or amount. It is essentially a variation of the bar chart in which a row of representative symbols is used to make up the bar.

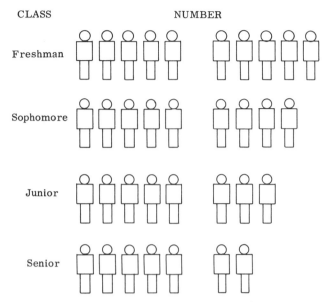

Each symbol represents 100 pupils

Histogram: A vertical bar graph in block form made by plotting frequency of occurrence against the values obtained. It is the common mode of representing frequency distribution in statistics (below).

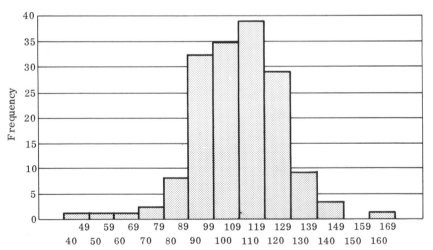

Circle graph (sectogram or pie chart): This graph is a pictorial representation of a complete circle, or "pie," that is sliced into a number of wedges, the size of each wedge showing its percentage of the entire quantity. This is the sort of chart you often get with your county tax bill to show you how your tax money is being spent. The example is shown at the top of the next page.

Where the Money Goes

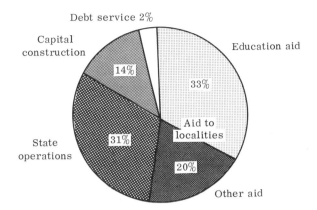

The other kinds of special graphs we will examine in Chapter 5 are conversion graphs, and a unique and very useful type of graph known as a nomograph, or nomogram.

What difference in appearance do you see between the histogram shown

on page 26 and the vertical bar graph shown in frame 22? _____

_____

- -

In the vertical bar graph the columns are separated, whereas in the histogram they adjoin one another. You will discover the reason for this difference later on when we examine them both in greater detail.

25. Since it will be helpful in your study of graphs if you can keep in mind the four principal categories into which we have divided them, see if you can write, in the spaces below, the names we have used for the four major types. It would be nice if you could put them in the order in which we have discussed them, but not absolutely essential.

(1) _____   (2) _____

(3) _____   (4) _____

- -

(1) line graphs; (2) surface graphs; (3) bar graphs; (4) special graphs

COORDINATE SYSTEMS AND PLOTTING PAPER

26. Having taken a brief look at the major types of graphs we will be studying, let's examine a little more closely several systems commonly used in plotting graphs.

First, let's review the grid system we referred to in frame 10. We said the grid system (also called a rectangular or Cartesian coordinate system) is a network of vertical and horizontal lines; but let's be more precise.

Going back to our graph showing the change of temperature with the month of the year, think for a moment about the fact that here are two quantities that change together. That is, time and temperature change together. Since in this graph the time and temperature vary or change, they are called <u>variables</u>. A graph, then, is a graphic representation of the relationship between two variables.

In the figure in frame 9, as we move along the horizontal "time-line" from March to April, the average value of the temperature increases 5°, from 42° to 47°. However, if we check the interval between October and November, the temperature decreases 8°, from 52° to 44°.

Let's take another example. This time we'll see how easy it is to make a graph or picture of a formula. To do this we will use the formula $d = 60t$, which tells us how many miles, d, a car will go in t hours if it travels 60 miles per hour. (We all use this formula in calculating driving time or distance. You may also recognize this relationship from an algebra course.) In this formula d and t are variables because they vary. The table below shows the distances gone in 0, 1, 2, 3...10 hours.

t	0	1	2	3	4	5	6	7	8	9	10
d	0	60	120	180	240	300	360	420	480	540	600

The graph of $d = 60t$ is shown at the right. Notice that in order to plot the pairs of values shown in the table we: (1) establish a network of evenly spaced vertical and horizontal lines; (2) establish a time scale along the horizontal base line; (3) establish a distance scale along the vertical base line; (4) locate (plot) points corresponding to the pairs of values shown in the table. The final step is, of course, to connect these points by a continuous line, giving us a straight-line graph.

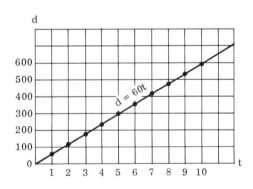

Now the really important thing to observe is that <u>this straight-line graph can be used to find the distance for any time shown on the graph, and to find the time for any distance shown.</u> To find the distance traveled in $5\frac{1}{2}$ hours, for example, we first find the point halfway between the 5 and 6 points on the horizontal base line (t-axis). Then with a straight-edge we find where the vertical line drawn from this halfway point cuts the plotted line. We then use the straight-edge to find where the horizontal line from this point on the graph cuts the vertical base line (d-axis). We estimate this point on the d-axis to be 330. In other words, when the time is $5\frac{1}{2}$ hours, the distance traveled is about 330 miles.

Using the graph, find out approximately how far the car would travel in

$8\frac{1}{2}$ hours. _____

- - - - - - - - - - - - - - - - - - - -

about 500 miles (Actually 510 miles. A larger graph would allow us to make
a more accurate measurement.)

27. Try using the graph in frame 26 the opposite way this time. That is, see
if you can convert a <u>distance value</u> to a <u>time value</u>. How long, according to

the graph, would it take for the car to travel 200 miles? _____

- - - - - - - - - - - - - - - - - - - -

A little under $3\frac{1}{2}$ hours. Starting along the vertical (distance) scale find 200
(miles). Follow the 200-mile horizontal line across to the right until it meets
the time-distance graph line. From this point on the graph drop straight down
(use a straight-edge to assist you if necessary) to the corresponding point below on the time scale. This point appears to be about halfway between the 3
and the 4, telling us that the time required to travel 200 miles at 60 miles per
hour is about $3\frac{1}{2}$ hours.

28. Most of us are familiar with this time-distance formula from planning auto
trips. But just in case you haven't had algebra let's review <u>why</u> d (distance)
and t (time) are variables.

   You will recall that the relationship between d and t given us in frame 26
was d = 60t. Suppose we wish to find the values for d as we substitute different values for t in the above formula (in other words, as we let t <u>vary</u>). With
a little quick mental multiplication you can see that if we assign a value of 1
(one) to t, then d will equal 60, since $1 \times 60 = 60$. Or if we are interested in
knowing how far the car will travel in 2 hours, we substitute a value of 2 for
t in the formula. We find thus that $d = 2 \times 60 = 120$, that is, that d = 120 miles.
Similarly, if t = 3, d = 180; if t = 6, d = 360, and so on. This is, of course,
the way in which the pairs of values in the table in frame 26 were found. In
turn our time-distance graph was obtained by plotting these values.

   Remember, <u>the primary function of a graph of any kind is to illustrate the
relationship between two variables</u>, such as d and t. Graphs are used to show
relationships between an almost limitless variety of variables in all scientific
and economic fields.

   To draw any graph we <u>must</u> have established some relationship between
the two variables. This relationship can be in the form of a formula (<u>equation</u>
is the more mathematical term), as we have just seen, or simply a set of observations, as is common in all types of statistical work. Sometimes we develop a set of observations and then try to <u>find</u> an equation that expresses, in
mathematical language, the relationship between the two variables. We will
delve into this a bit more deeply when we study the specific types of graphs.

(a) In the formula p = 25k, what are the variables? _____

(b) In the formula p = 25k, if k equals 4 what does p equal? _____

(c) What is the principal purpose of a graph? _____

_____

- -

(a) p and k; (b) 100; (c) to show the relationship between two variables
    If you got answer (a) correct, you <u>inferred</u> it from the previous formula,
d = 60t. And quite properly, since this is all you had to go on. However, it
illustrates something you may already be aware of (especially if you have studied algebra), namely, that letters of the alphabet usually are used to represent
variable quantities.

29. The relationship d = 60t is an example of what we call <u>direct variation</u>.
That is, as t increases, d likewise increases linearly. You can see this from
both the graph and the table in frame 26. However, not <u>all</u> variation is direct
variation. Sometimes as one variable increases the other variable decreases.
This relationship is called <u>inverse variation</u>.
    For example, consider this more general formula for the time-speed-
distance relationship: $s = \dfrac{d}{t}$, or speed $= \dfrac{\text{distance}}{\text{time}}$. Since this formula contains
<u>three</u> variables (one more than we are interested in at the moment) we will as-
sign a value of 100 (that is, 100 miles) to d, which it is perfectly proper to do,
mathematically. This makes it a <u>constant</u> instead of a variable and narrows
down the field to the two variables we are interested in, that is, s and t. Here
is the formula as we now have it: $s = \dfrac{100}{t}$.
    What do you think will happen to values of s as the values of t increase?

_____

- -

As the values of t increase, the values of s will decrease. (If you're not sure
why, read on.)

30. Let's return to the formula $s = \dfrac{100}{t}$ and find out why s decreases as t increases. The best way to do this will be to substitute a few increasing values
for t and see what happens to s.
Doing so gives us the values
shown in the table at the right.
    As the table indicates, if t

t	1	2	4	5	8	10
s	100	50	25	20	$12\frac{1}{2}$	10

is 1 (one hour), s becomes 100 (mph). If t is 2, $s = \dfrac{100}{2}$ or 50 mph. When
$t = 4$, $s = \dfrac{100}{4} = 25$ mph, and so on. Plotting these values gives us the curve
shown at the top of the next page. This tells us that whereas direct variation
results in a straight-line plot, inverse variation yields a curvilinear graph.

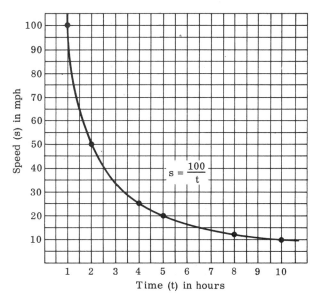

Summarizing, and substituting the letter k for the constant in each of our two examples:

Direct Variation: $d = kt$, or $k = \dfrac{d}{t}$, hence the quotient of the two variables is constant and either variable increases when the other increases; the plot is a straight line.

Inverse Variation: $s = \dfrac{k}{t}$, or $k = st$, hence the product of the two variables is constant and either variable decreases as the other increases; the plot is a curved line.

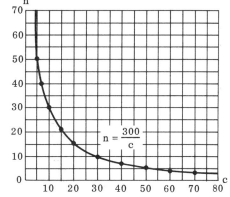

Obviously, therefore, when looking at a graph showing the relationship between two (first degree) variables you can tell at once if the relationship is direct or inverse simply by observing the shape of the plotted line.

Without even knowing what the variables are in the graph at the right, would you say that they vary directly or inversely with one another?

— — — — — — — — — — — — — — — — — — —

inversely

31. Now let's take a closer look at the method of plotting points on the grid system. So far all the graphs we have seen represented the relationship of two variables when both of them had positive values. However, often we wish

to picture the relationship between two variables when either or both of them has <u>negative</u> values.

The graph at the right, for example, shows the relation of centigrade and Fahrenheit (temperature) readings only when both readings are positive. Can you think of a way to show the relationship of C to F where F is less than 32°?

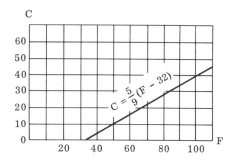

- - - - - - - - - - - - - - - - - - - -

The best answer is to extend each axis (the vertical and horizontal base lines) in both directions from the origin, 0. We then would have a chart that looked something like the one shown at the right.

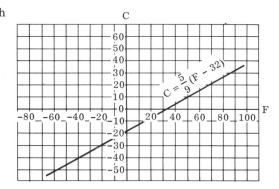

32. By extending each axis in both directions from the origin, 0, we can locate any point on the paper (or <u>in the plane</u>, as a mathematician would say) if we know its distance and direction from each axis.

Think of the two axes as two principal streets which meet at right angles at the center of a small town, where there is an information booth. If a stranger at the origin O wishes to find a house, A, of historic interest, he can be directed as follows: "Go three blocks east and two blocks north." To find the church at point B he would go 3 blocks west and 4 blocks north; to find the ball park at point C he would go 5 blocks west and 2 blocks south; and to find the motel at point D he would go 6 blocks east and one block south.

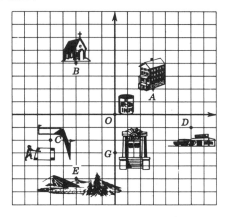

If we call the distance east positive, then we should call the distance west negative. And if the distance north is positive, then the distance south is negative. So instead of saying "two blocks north" we could say "+2"; instead of saying "two blocks west" we could say "-2"; and instead of saying "two blocks south" we could say "-2."

In stating the location of a point we always give the horizontal distance first. Thus, "two blocks east and 3 blocks south" is written (+2,-3), or just (2,-3) since a number is considered to be plus (positive) unless otherwise marked.

If you have studied some mathematics you'll remember that the horizontal distance is called the <u>abscissa</u> and the vertical distance the <u>ordinate</u>. The abcissa and ordinate of a point are known as the <u>coordinates</u> of the point. In the diagram at the bottom of page 32, the coordinates of A are (3,2); of B are (-3,4); of G are (0,3); and of E are (-3,5).

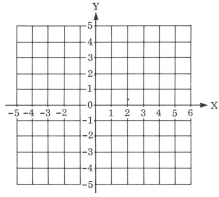

Using the coordinate system at the right see if you can locate the points listed below.

(a) (4,1)   (b) (-3,3)   (c) (-5,-4)   (d) (5,-2)

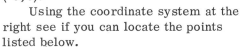

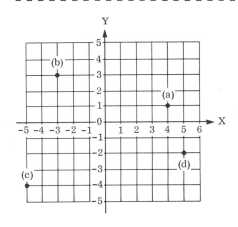

You will notice that we labeled the horizontal axis the X-axis, and the vertical axis the Y-axis. This is conventional in mathematics simply because when we are working with two variables that represent purely numerical values—or whose identity we don't know—we identify them just as x and y. Therefore in plotting points in the graph showing the relationship between these two variables we are plotting (x,y) values, hence we name our coordinate axes accordingly.

A relatively small amount of our work with graphs will involve the use of mathematical equations. But we will have some of this to do and it is important, therefore, that you be familiar with these concepts. Incidentally, there are coordinate systems other than rectangular. There also are several kinds of special plotting paper in addition to rectangular graph paper with evenly spaced lines.

TYPES OF LINEAR RELATIONSHIPS

33. Each type of graph has some unique advantages. It may also have some disadvantages, depending upon the use you have in mind for it. To choose and use charts effectively you must learn to select the best graph to suit a particular need. Frequently several modes of presentation will do. The important question is, Which will be best? So let's consider what appear to be the most useful characteristics of the four major categories. You need to be aware of these characteristics both as a chart user and a potential chart maker.

Line graphs, either singly or in combination with other types of graphs, probably appear more frequently than any other kind. They can be used for the simplest comparisons or the most complex. They are extremely flexible and can be adapted to a wide range of uses. They are compact, easily carrying much more data than any other type of graph. They are easier to draw than other types.

Basically, line graphs serve to show the relationship between two variables, using rectangular coordinates. We saw this in the time–speed–distance graphs. They really may be the most basic and essential sort of graph, for not only can they be used to display information but—even more importantly—they also can be used to solve problems. Sometimes two graphs have to be studied together to get the required information, but often this job is simplified by drawing two lines on one graph.

Suppose, for example, we wish to compare the performances of two cars. One car (driven by an "average" driver) travels at an average speed of 40 mph. The other car (driven by the proverbial "little old lady") travels at an average speed of 20 mph. We would like to get some idea of how long it is going to take for these two vehicles to cover a comparable distance. Recalling our basic time–speed-distance formula from frame 29,

$s = \dfrac{d}{t}$, we solve (restate) it for d, giving us the relationship d = st, that is, distance equals speed multiplied by time (or vice versa). Using the known speeds of the two cars and substituting various values for time (t) gives us this table of values (shown at the right) which, when plotted, produces the two straight–line graphs shown at the top of the next page.

Elapsed time (hours)	Distance traveled	
	Car A	Car B
1	40	20
2	80	40
3	120	60
4	160	80
5	200	100
6		120
7		140
8		

(a) Use the information in the table to plot the missing 5-hour and 7-hour distance points for cars A and B respectively, on the graph.

Use the graph to answer the following questions:

(b) How far will car A go in $4\frac{1}{2}$ hours? _____

(c) How long will it take car B to go 50 miles? _____

(d) How does car A's speed compare with car B's speed?

_____

_____

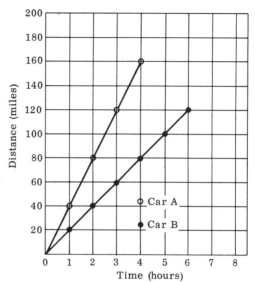

- -

(a) (See graph at right.)
(b) 180 miles
(c) $2\frac{1}{2}$ hours
(d) A's speed is twice that of B's.

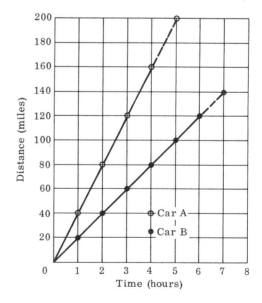

34. Although the type of dual or compound graph shown in the last frame is quite a simple one, it can be used to show the relationship between two variables, convert time values to distance values and vice versa for two different car speeds, and compare, both visually and numerically, the different rates of change of two quantities. In the example we considered that these rates represented speeds of cars but, using different units, the graph might have represented many quantities varying as constant rates. We could, for instance, use a compound graph of this type to show the relative rates at which two

different chemicals dissolved. The graph would then represent rates that were constant but different for the two chemicals. That is, each chemical dissolved at a steady rate over a period of time but each had its own rate.

Such rates are not always constant, however. Suppose we wanted to draw a graph showing the rate at which some chemical dissolved but the rate of dissolving was <u>not</u> constant. In other words, it dissolved most rapidly when first put into water but then gradually, but steadily, slowed down in its rate of dissolving.

What do you think the graph showing the solution of the chemical in water over a period of time would look like? _____

- - - - - - - - - - - - - - - - - - - -

It would be curvilinear, <u>not</u> a straight line. The graph would be a straight line <u>only</u> if the rate of solution were constant.

35. Let's look at such a situation. We will assume we have some way of measuring, instantaneously, the amount of chemical in solution at intervals of 10 seconds after the dry chemical has been added to a certain amount of water. Below is the table of observations you have recorded. Plot them on the coordinate system provided and see what the resulting graph looks like.

Time (seconds)	Grams in solution
10	2.0
20	3.5
30	4.7
40	5.8
50	6.8
60	7.7
70	8.5

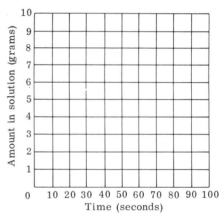

Note: In plotting these values you will have to make the best estimate you can of the decimal fraction values between the whole numbers (grams) shown along the horizontal axis of the chart.

- - - - - - - - - - - - - - - - - - - -

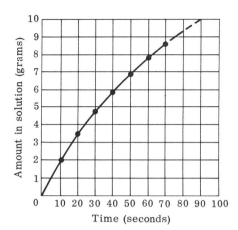

36. What do you suppose the graph would look like if the chemical had dissolved at a uniform rate of 2 grams every 10 seconds? See if you can work it out. At the right is a partially completed table showing the time values, just as in the previous frame. However it will be up to you to fill in the corresponding values for the number of grams in solution at each 10-second interval. When you have done so, plot your new table of values on the graph in frame 35 so that you can compare the two curves.

Time (seconds)	Grams in solution
10	2
20	____
30	____
40	____
50	____
60	____

- -

Time (seconds)	Grams in solution
10	2
20	4
30	6
40	8
50	10
60	12

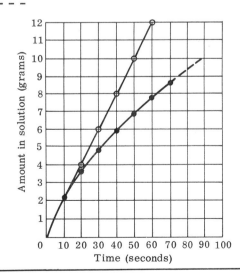

37. In this last example we have two different types of line graph combined in the same chart—one a straight line and the other curvilinear. Notice how nicely this helps us visualize what happens in two related but different situations. The variables (time and grams of chemical in solution) are the same in both cases, but the <u>relationship</u> between them differs.

It is only fair to point out that in drawing the curve shown in frames 35 and 36 we used a drafting aid known as a "French curve." This raises a point of which you should be aware, either when drawing or simply using curvilinear graphs: Plotted points don't always fall in such a way as to form the structure of a smooth curve. That is, they're not always so well lined up. Sometimes, for example, four out of a total of six points form a smooth pattern, but the other two points simply don't fit this pattern—not exactly, at least.

In linking up points in a graph that do not follow a regular pattern we generally are attempting to show one of three things: (1) an observed relationship (like the temperature readings we worked with earlier); (2) an empirical relationship; or (3) a theoretical relationship. To understand what this means, consider the following examples.

The graph at the right is made to show an <u>observed relationship</u>, giving a point-by-point plot of the data obtained by experiment or investigation. Note that the points have simply been connected with straight-line segments, making this a zigzag curve. The resulting curve is only one way of deriving a graphic portrayal of the relationship between the variables, based on observation.

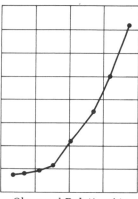

Observed Relationship

This second graph shows the same set of points, also based on observation, but the smooth curve shown represents the investigator's <u>interpretation</u> of his series of observations. The smoothed curve has been fitted to the data by eye—or by a formula chosen empirically.

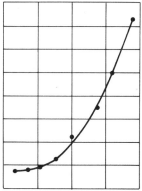

Empirical Relationship

This third graph is of a curve derived from <u>theoretical considerations</u> alone, with the observed points plotted to show consistency of observation with theory.

The foregoing examples relate to what is termed "curve fitting," the problem of finding a curve that best fits the data. The curve shown in the graph below represents the plot of a number of observed data points, with a smooth curve fitted to the data by eye.

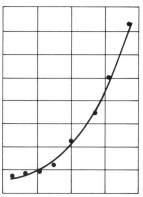

Theoretical Relationship

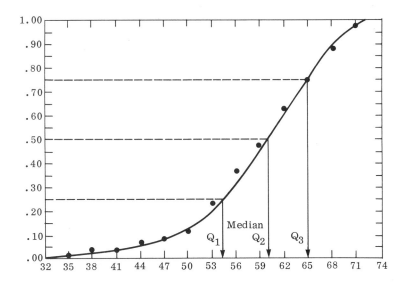

From what we have just discussed, what would you call the relationship pictured in the graph at the bottom of the previous page? _____

_____

– –

an empirical relationship

38. From what we have learned so far about graphs we should be able to draw some conclusions regarding their advantages and disadvantages. Let's state these now so that we will have them in mind during the more detailed discussion of graph interpretation that follows. <u>The type of graph selected to do a particular job should be one that maximizes the advantages and minimizes the disadvantages.</u>

<u>Pros and Cons of Graphs</u>

<u>Advantages</u>
Quick: Show the main features of the data at a glance.
Forceful: Carry much more emphasis than text or tables.
Convincing: Prove the point instead of merely stating it.
Compact: Pack a lot of information in a small space.
Interesting: Easier to look at than text or tables.

<u>Disadvantages</u>
Technical: Some readers unfamiliar with the interpretation of graphs.
Demanding: Require special know-how to design effective graphs.
Costly: Take more time to construct.
Not always usable: Some data not suitable for graphic representation.
Not precise: Cannot be read as accurately as text or tables.

Which of the five advantages listed above would you say the graph at the right has?

_____

_____

_____

_____

_____

– – – – – – – – – – – – – –

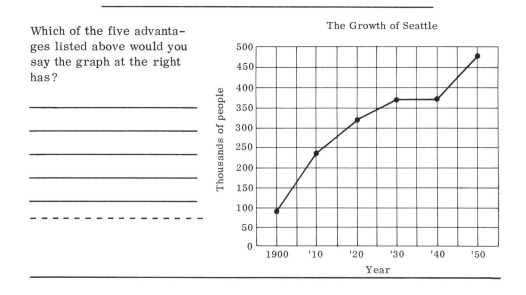

The Growth of Seattle

I'd say it has all of them. The main feature of the data (population growth of Seattle) certainly is apparent at once. And the simplicity and direction of the line curve could hardly carry the message more forcefully or convincingly. The compactness of the information is obvious. Whether or not one finds this graph more interesting to look at than the same data presented in text or tabular form is, prehaps, a matter of opinion. But considering that most people have difficulty in extracting information from a textbook this chart probably is a clearer mode of communication than either a text or table—although a table would certainly be a close second choice.

FROM TABLE TO GRAPH

39. Having given some attention to the logical development of data relationships in tabular form and then in graphic form, and having as well considered some of the ways in which graphs are developed and used, let's look more carefully into the process of developing graphs from tabular information. We've done some of this before (the temperature chart and the time-speed-distance graph), but there is more to learn.

A well-thought-out table sometimes holds so much information or deals with a subject so important that you could make several useful graphs from it. But usually one or two meaningful charts is as much as you can expect from any one table. The problem is not how <u>many</u> charts can be made from these data, but whether or not a significant chart can be made, and if so, what is the best type to use.

A graph presents a limited number of figures in a bold and forceful manner. To do this it usually must omit a large number of figures available on the subject. <u>The choice of what graphic format to use is largely a matter of deciding what figures have the greatest significance to the intended reader and what figures he can best afford to skip</u>. If you are the intended reader (rather than the creator), knowing this will tell you something about what the chart-maker thought was important and so assist your interpretation.

If, on the other hand, you are the one faced with the problem of preparing a suitable graph, and if you know a potential reader is interested in a specific question, don't overlook it in making your choice. Study the data to see if you can find any other relationships or tendencies you think the reader should know. Consider how the key figures in the chart might be used as a basis for action, as an explanation, or as a justification.

After you have decided which data are the most important to graph there is still the question of what kind of chart to use. Even when you feel you know what type of chart to use it is sometimes wise to try others. Occasionally you will find that your first choice is not the best one. In fact, it may be necessary to sketch several types before you find one that focuses most clearly on the facts you want to bring out.

In the next few frames we're going to examine a data-source table and the graphs developed from it. In doing so we're borrowing an example taken directly from Army administrative records showing the number of general purpose vehicles disabled four days or more over a period of six months. The subject matter may not sound very exciting but I believe you will be interested in seeing the eight different types of charts developed to display the tabular information in a more impactive form.

These eight charts are not meant to suggest that you should try to get several graphs from every set of data. They are shown only to prove that the facts in even a simple summary table can be brought out in several different ways.

Notice as you come to them that no two of the charts are exactly the same. Each one is different because its <u>purpose</u> is different. Each focuses attention on a different aspect of the data—or puts a different emphasis on a given aspect. This is the most important point to remember whenever you design a chart: <u>When you change the chart you change the story</u>. The choice of what type of chart to use is not a matter of taste; it is always a matter of judgment.

Service	May		June		July		August		September		October	
	Total	Due to lack of parts	Total	Due to lack of parts	Total	Due to lack of parts	Total	Due to lack of parts	Total	Due to lack of parts	Total	Due to lack of parts
(1)	(2)	(3)	(4)	(5)	(6)	(7)	(8)	(9)	(10)	(11)	(12)	(13)
All Services	922	625	1,271	774	856	533	981	675	1,247	679	1,486	682
Chemical	98	48	78	43	78	23	39	17	91	51	95	51
Engineers	136	116	136	97	29	29	19	18	29	29	21	17
Medical	41	25	81	44	4	2	6	5	5	2	9	6
Ordnance	252	174	368	221	265	193	353	268	606	218	654	164
Quartermaster	116	63	76	41	61	21	61	23	59	31	58	28
Signal	69	44	80	34	28	15	31	23	39	34	46	30
Transportation	210	155	452	294	391	250	472	321	418	314	603	386

What are the four main items of information the above table gives us?

_____

_____

_____

_____

- -

(1) The total number of general purpose vehicles for <u>all</u> services disabled four days or more in each of six months. (2) The number of general purpose vehicles for <u>each branch</u> of the service disabled four days or more each month.
(3) The total number of general purpose vehicles for <u>all</u> services disabled four days or more each month due to lack of parts. (4) The number of general purpose vehicles for <u>each branch</u> disabled four days or more each month due to lack of parts.

40. The next four frames (40–43) show how the data from the table in frame 39 may be presented to emphasize <u>changes that have occurred</u> from one period to the next during the entire time span covered by the table. They not only show how things stand at present but how they stood at earlier periods and, to some extent, provide a basis for estimating where they may be heading.

Chart A puts the main emphasis on the total number of vehicles on which repair was delayed four days or more. It shows also the composition of the total by cause. This chart stresses the changes in the total from period to period and also the changes in the component next to the base line. Changes in the other component are not as easily read and direct comparisons between the two components are not provided.

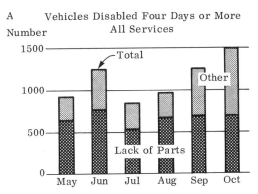

A   Vehicles Disabled Four Days or More
All Services

(a) In chart A, what is the name of the component next to the base line?

_____

(b) What is the name of the component occupying the upper portion of the vertical bars? _____

(c) Was there any month in which lack of parts was not the main reason for vehicle disablement? If so, which month? _____

(d) Do you see a pattern or trend developing in the total number of vehicles disabled? If so, during what months does the trend occur and what does it indicate? _____

_____

- - - - - - - - - - - - - - - - - - - -

(a) Lack of Parts; (b) Other (i.e., factors other than lack of spare parts); (c) Yes. October, when reasons other than a lack of spare parts was the major cause of vehicle disablement. (d) Yes. From July to October the trend in total number of vehicles is consistently up, increasing primarily for reasons <u>other</u> than lack of parts.

41. Chart B (at the top of the next page) shows the same information as chart A but is designed to compare the components directly. It also provides a comparison of each component and the total. The composition of the total is not readily apparent but is easy to see if you look for it. It can be read with either component at the base. This graph provides a better picture than chart A of the interrelationships of the data.

(a)   What was the total number of vehicles disabled during the month of August (approximately?

_____

(b)   How do you suppose the chartmaker found this total (for August)?

_____

_____

_____

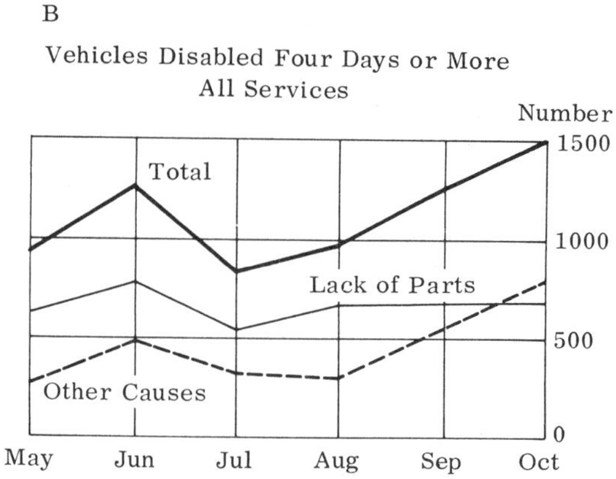

B

Vehicles Disabled Four Days or More
All Services

(c)   Does this graph show the same trend you (hopefully) observed in frame 40? _____

(d)   Do the Total and Lack of Parts curves always move together (that is, both up and down each month)? _____

- -

(a) 1000; (b) By adding the two components that comprise it: Lack of Parts (about 700) and Other Causes (about 300). (c) Yes, even more clearly (upward, July through October). (d) No. During the period of August to October the Total moves upward while the Lack of Parts line remains horizontal.

42. Chart C shows the relative importance of the lack of spare parts as a cause of delay. This picture is not quickly apparent in chart A but is somewhat more so in chart B. This type of chart provides a picture of changes in a significant ratio, without regard to the absolute values behind the ratio.

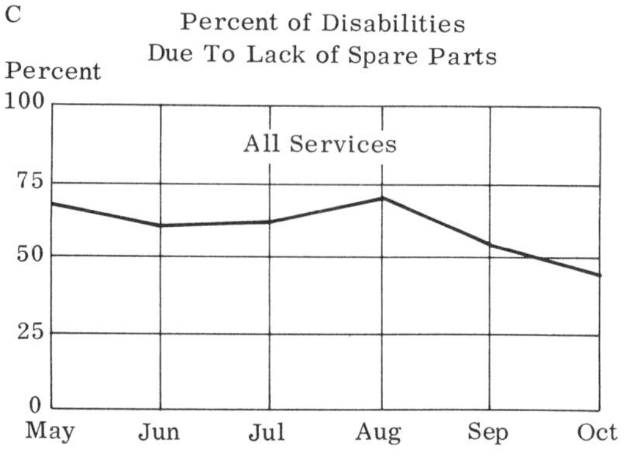

C

Percent of Disabilities
Due To Lack of Spare Parts

(a)   On this chart, what would a value of 100% indicate—assuming that the curve touched this value?

_____

(b)   In what month did disabilities due to lack of spare parts fall below 50%? _____ Does this answer agree with your answer to question (c) in frame 40? _____

- -

(a) Vehicle disabilities (during that particular month) were due <u>entirely</u> to a lack of spare parts; (b) October. It should, since both questions bring out the same point.

43. Chart D shows the same data as chart C, but it also shows the range between the high and low service for the branches each month. This presentation amplifies the picture shown in chart C, bringing out the spread among the figures from which the "All Service" values are derived.

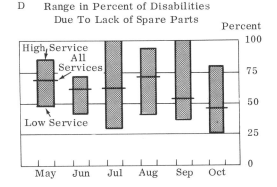

D   Range in Percent of Disabilities Due To Lack of Spare Parts

(a) According to this chart what is the <u>range</u> (in percent) of disabilities due to lack of spare parts between the high service (i.e., the service branch with the highest figure) and the low service (branch with the lowest figure) during the month of October?

_____

(b) What is the All Services figure for August? _____

(c) What is the lowest Low Service percentage figure shown here?

_____ In what month did it occur? _____

- - - - - - - - - - - - - - - - - - - -

(a) about 53%—that is, from 25% to about 78%; (b) about 74%; (c) 25%, October

44. Now we're going to take a look at four charts that emphasize the <u>relationships between components</u> of the total for a specific period. Two of the graphs examine the causes of disability, and two focus on the proportion of vehicles disabled due to lack of spare parts. The first two present a cross-section of the total at a specific period of time. The third provides a secondary comparison that measures one period against another. The fourth summarizes results during the entire span covered.

    The first of these charts, chart E, shows for each technical service the <u>total number</u> of vehicles disabled four days or more, and also shows how many were delayed in repair by lack of spare parts and by other causes. The main comparison here is between totals. However, putting the "Lack of Parts" segments next to the base line permits this

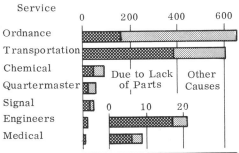

E   Number of Vehicles Disabled Four Days or More—October 1971

secondary comparison. (The inset graph is an expansion scale for the two lowest services.)

(a) About how many vehicles of the Transportation service were disabled four days or more due to lack of parts during the month of October?

_____

(b) How many Ordnance vehicle disabilities were due to other causes?

_____

(c) How many Engineers vehicles were disabled due to lack of parts?

_____

- -

(a) something less than 400; (b) about 500 (650 - 150); (c) about 18

45. Chart F shows <u>the proportionate part</u> of delays in repair that resulted from lack of spare parts and from other causes. Since this type of presentation measures the relative importance of each component, it often reveals facts not readily apparent in a chart showing absolute values.

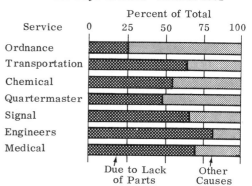

F Percent Distribution of Vehicles Disabled Four Days or More—October 1971

(a) According to this chart, what <u>percentage</u> of the vehicles of the Chemical service were disabled four days or more during the month of October due to causes <u>other</u> than lack of spare parts?

_____

(b) For which branch of the service was a lack of spare parts highest as the cause of vehicle disablement during October? _____

What was the percentage? _____

(c) For which branch of the service were "Other Causes" greatest as the cause of vehicle disablement? _____

- -

(a) approximately 45%; (b) Engineers, about 80%; (c) Ordnance

46. Chart G (at the top of the next page) emphasizes the comparison of the "Lack of Parts" percentages shown at the left of the chart. Although the "Other Causes" segment is not shown, it can readily be seen as the difference between the bars and the 100-percent line. As an added feature this graph

also shows the same measurement for an earlier period, thus permitting a three-way comparison of current results, earlier results, and change from one time to the other.

Percent of Vehicles Disabled Four Days
or More Due to Lack of Parts
G   September and October

(a) How does the percentage of Signal Corps service vehicles disabled during October due to lack of parts compare with the percentage during September?

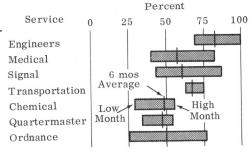

(b) What percentage of vehicles was disabled during September in the Medical Corps? _____

(c) Which branch of the service had a disability rate of 75% for September due to lack of parts? _____

(d) Which service had the <u>lowest</u> rate for October due to lack of parts?

_____

- - - - - - - - - - - - - - - - - - - -

(a) about 20% less; (b) about 37%; (c) Transportation; (d) Ordnance

47. Chart H carries the story presented in chart G a step further by showing an overall picture for the entire period covered instead of just the current period. Emphasis is on the range in the "Lack of Parts" percentage during the period, without identifying specific months.

H
Percent of Vehicles Disabled Four Days
or More Due to Lack of Parts
Range for Period May–Oct 1971

(a) What is the percentage <u>range</u> of Ordnance vehicles disabled four days or more due to lack of spare parts for the period May to October?

_____

(b) What was the 6-month average for the Transportation branch?

_____

(c) Which service had the lowest range-end value? _____

What was this percentage value? _____

(d)   What was the high month value for the Signal Corps? _____

- -

(a) about 52%—from 25% to about 77%;   (b) around 65%;   (c) Ordnance, 25%
(d) a little under 90%

    Having taken an overall look at the uses of tables and graphs, some of the most common fallacies of statistical presentation, the major categories of graphs, types of coordinate systems, the pros and cons of graphs, and the relation of tables to graphs, and having developed some facility in interpreting graphs and tables, it is time now for a short review test to help you find out how you are getting along.

    Be sure to review the material in the chapter as needed, based on the results of this test. Having arrived at the end of Chapter 1 you are now ready to investigate in greater detail the types of graph that interest you most or will be most useful to you. The remaining chapters are arranged to allow you to choose what best suits your own purposes. The chart in the Appendix might help you locate the material you need.

SELF-TEST

1.   What effect does scale expansion or scale shrinkage have on the message a curve conveys to the reader? _____

_____

2.   Find the following averages of the column
of numbers at the right.

(a) Mean _____

(b) Median _____

(c) Mode _____

85
80
80
80
70
70
30
25
20
5
5

3.   In 1952 the Chicago Police Department prohibited the showing of the Italian film, "The Miracle." An interested citizens organization reported the following investigation.

        In the past few months our Censorship Committee has shown
        "The Miracle" at several private meetings. Of those filling
        out questionnaires after seeing the film, less than 1 percent
        felt it should be banned. "It would seem," said the Chairman

of the Censorship Committee, "that the five members of the Police Department's Censorship Board do not represent the thinking of the majority of Chicago citizens."

What possible fallacies, if any, do you see in this report? _____

_____

_____

4. The illustration at the right showing a pair of blast furnaces was intended to illustrate how the iron and steel industry's steelmaking capacity boomed between the 1930's and the 1940's. The blast furnace representing the ten-million-ton capacity added in the '30's was drawn just two-thirds as tall as the one for the fourteen and a quarter million tons added in the '40's.

Do you feel this conveys the correct impression as to the relative increase in capacity in each

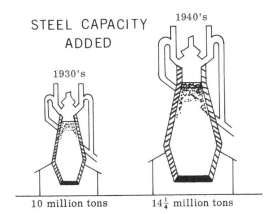

STEEL CAPACITY ADDED

1940's

1930's

10 million tons   14$\frac{1}{4}$ million tons

of the two decades? Explain your answer. _____

_____

5. How would you go about finding the arithmetic average of a group of numbers? _____

_____

_____

_____

6. In the figure at the right:

(a) What are the values in the vertical scale? _____

_____

(b) What are the values in the horizontal scale? _____

_____

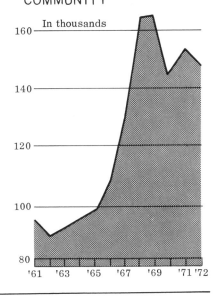

EMPLOYED IN EXCHANGE COMMUNITY

In thousands

160

140

120

100

80

'61   '63   '65   '67   '69   '71 '72

(c)   How much (that is, how many units) is the space between any two horizontal lines equal to? _____

7.   The vertical and horizontal lines of a rectangular coordinate system are at _____ to one another.

8.   State at least one advantage a line graph has over a tabular presentation of data. _____

_____

9.   Suppose you have collected data on a group of women regarding their satisfaction with a household product. The women range in age from 20 to 50 years. You want to arrange your data according to five-year age groups and degree of satisfaction or dissatisfaction in the following categories: very satisfied, satisfied, neutral, dissatisfied, very dissatisfied. Draw a sketch showing what your tabular form would look like.

10.   Let us suppose that the data below are the results of your survey in question 9.

> Very Satisfied: age 20-25—12; age 26-30—9; age 31- 35—7; age 36-40—6; age 41-45—5; age 46-50—5
>
> Satisfied: age 20-25—9; age 26-30—9; age 31-35—7; age 36-40—4; age 41-45—3; age 46-50—2
>
> Neutral: age 20-25—5; age 26-30—4; age 31-35—6; age 36-40—5; age 41-45—4; age 46-50—1
>
> Dissatisfied: age 20-25—2; age 26-30—1; age 31-35—4; age 36-40—4; age 41-45—5; age 46-50—7
>
> Very Dissatisfied: age 20-25—0; age 26-30—1; age 31-35—3; age 36-40—6; age 41-45—8; age 46-50—11

Enter this data on the table you sketched in question 9. (If you still needed convincing, this certainly demonstrates the usefulness of tables over text!)

What trends, if any, do you see in it? _____

_____

_____

_____

11. Now, in order to understand better the nature of the
relationship between the sets
of variables shown in the answers to questions 9 and 10,
use the section of graph paper
at the right (or any other) to
plot the data points for the
"Very Satisfied" consumers
versus the six age groups,
again using the data from the
table in answer 9. Let the age
groups appear along the hori-

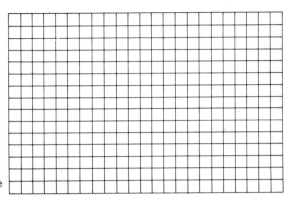

zontal scale and the number of "Very Satisfied" customers along the vertical
scale. (Cover the answer to this question while using the table in answer 9
in order to test fairly your ability to plot such data.)

12. The table below shows tractor fatality rates on the farm by type of accident for the period 1969-1972. Referring to the two 1969-1972 Average columns in the right half of the chart, answer the questions which follow.

**Tractor Fatality Rates on the Farm
by Type of Accident, 1969-1972**

Type of Accident	Average Rate*		1969-72 Average	
	1969-70	1971-72	Rate*	Per Cent
Total..............	16.3	15.8	16.0	100%
Overturns.............	8.5	8.6	8.6	53
Run over.:............	4.8	3.5	4.1	26
Power takeoff.........	0.6	0.9	0.7	4
Other.................	2.4	2.8	2.6	17

Source: Deaths based on reports from 10 state vital statistics registrars and individual case reports; tractors based
on U. S. Census of Agriculture, 1969.
*Number of fatalities per 100,000 tractors.

(a) What type of accident had the highest fatality rate? _____

(b) What percentage of total on-the-farm tractor fatalities did run-overs ac-

count for? _____

13. The table below represents a breakdown of drivers by age group and shows
the number of accidents—combined and fatal—each group was involved in.

Age of Drivers —
Total Number and Number in Accidents, 1972

Age Group	All Drivers		Drivers in Accidents					
	Number	%	Fatal		All		Per No. of Drivers	
			Number	%	Number	%	Fatal*	All**
Total..............	118,200,000	100.0%	70,900	100.0%	29,100,000	100.0%	60	25
Under 20..........	12,200,000	10.3	11,500	16.2	5,200,000	17.9	94	43
20-24.............	13,300,000	11.3	13,700	19.3	5,400,000	18.6	103	41
25-29.............	12,800,000	10.8	8,300	11.7	3,600,000	12.4	65	28
30-34.............	11,600,000	9.8	7,200	10.2	2,800,000	9.6	62	24
35-39.............	11,200,000	9.5	5,200	7.3	2,350,000	8.1	46	21
40-44.............	11,300,000	9.6	5,200	7.3	2,100,000	7.2	46	19
45-49.............	11,300,000	9.6	4,800	6.8	2,100,000	7.2	42	19
50-54.............	10,300,000	8.7	3,800	5.4	1,600,000	5.5	37	16
55-59.............	7,900,000	6.7	3,100	4.4	1,400,000	4.8	39	18
60-64.............	6,200,000	5.2	2,700	3.8	1,000,000	3.4	44	16
65-69.............	4,600,000	3.9	2,000	2.8	850,000	2.9	43	18
70-74.............	3,200,000	2.7	1,500	2.1	300,000	1.0	47	9
75 and over.......	2,300,000	1.9	1,900	2.7	400,000	1.4	83	17

Source: Drivers in accidents based on reports from 22 state traffic authorities. Number of drivers by age are NSC
estimates based on reports from state traffic authorities and research groups.
*Drivers in Fatal Accidents per 100,000 drivers in each age group.
**Drivers in All Accidents per 100 drivers in each age group.

(a)   Which age group was involved in the most fatal accidents?

_____

(b)   Which age group was involved in the most accidents of all kinds?

_____

(c)   Which age group would you expect (based on this chart only) to have the highest auto insurance rates?   _____

14. In which category would you place the graph at the right: straight-line, curvilinear, zigzag, or step?

_____

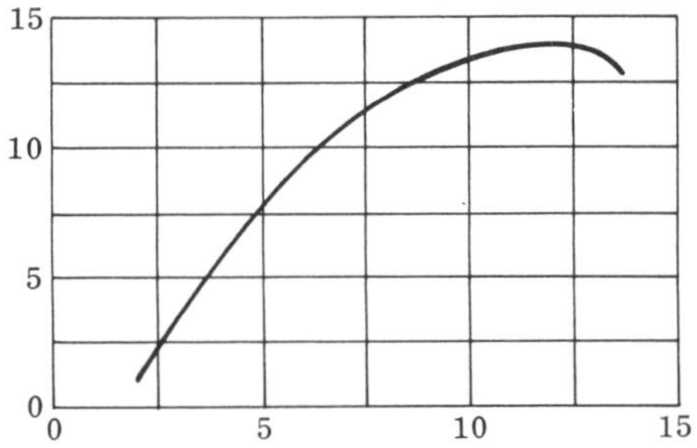

15. In a surface graph we are interested primarily in (the curve/the area under the curve). _____

16. The graph at the right is known as a _____

_____ .

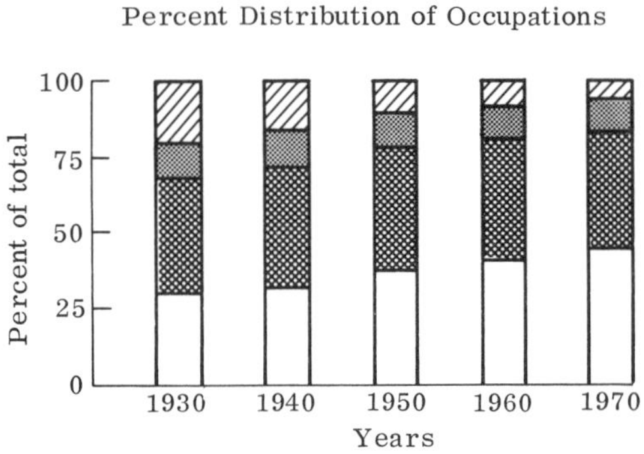

Percent Distribution of Occupations

17. This kind of chart is known as a _____

_____ chart.

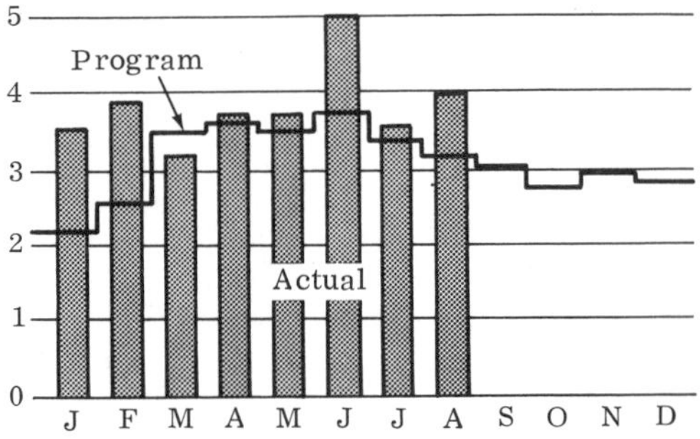

18. Identify the three types of chart shown here as either pictograph, pie chart, or histogram.

(a)

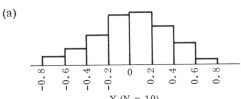

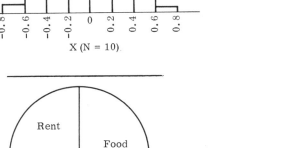

X (N = 10)

(a) _____

(b)

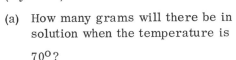

(b) _____

(c)

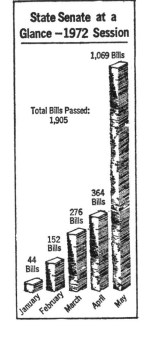

(c) _____

19. Name the four principal categories of graphs and charts. _____

---

20. At the right is the graph showing the number of grams in solution, at different temperatures, of a certain (mythical) chemical.

(a) How many grams will there be in solution when the temperature is 70°? _____

(b) At what temperature will there be 45 grams in solution? _____

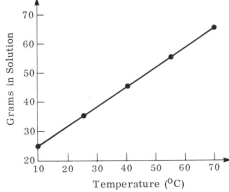

(c) From the appearance of the curve would you say that the relationship between the number of grams in solution and the temperature seems to be linear? _____

21. In the formula $y = 15x$, what are the variables? _____

22. In the formula $y = \dfrac{50}{x}$, what will happen to the value of y as x <u>decreases</u>

in value? _____

23. In the space at the right draw a rectangular coordinate system having a Y-axis and an X-axis, mark a scale along each axis, and locate the points listed below.

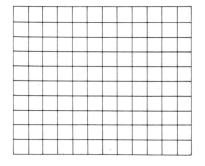

(a) (4, 2)
(b) (-4, -2)
(c) (-4, 2)
(d) (4, -2)

24. The graph at the right represents the time-distance relationship for two cars: A, traveling 50 mph, and B, traveling 30 mph. Find:

(a) How far car A will go in $2\frac{1}{2}$ hours.

_____

(b) How long it will take car B to go

150 miles. _____

Show the working lines you draw to assist you.

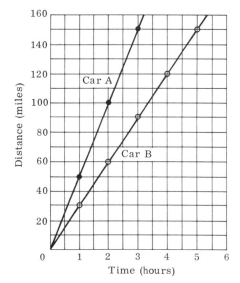

25. Use the coordinate system at the right to draw the graph of the table of x and y values (that is, values of the variables) given. Draw as smooth a curve as you can through the points you plot.

x	y
1	1
2	4
3	9
4	16

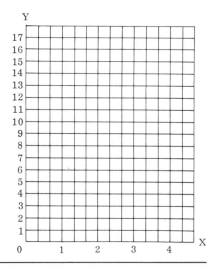

26. A graph showing a point-by-point plot of data obtained from an experiment or investigation, with the points connected by short, straight lines is termed

an (observed, empirical, theoretical) _____ relationship.

27. The table at the right relates to estimated cut in income tax for a married couple with two children as a result of special legislation.

(a) If your gross income is $10,000, your 1974 tax before rebate will be

_____.

(b) If your gross income is $15,000, your 1974 tax after rebate will be

_____.

(c) If your gross income is $15,000, your 1975 tax will be

_____.

MARRIED COUPLE—TWO CHILDREN			
Gross Income	1974 Tax Before Rebate	1974 Tax After Rebate	1975 Tax
$ 5,000	$ 102	$ 90	$ 0
$ 7,500	$ 488	$ 429	$ 163
$ 10,000	$ 867	$ 763	$ 518
$ 12,500	$ 1,261	$ 1,110	$ 961
$ 15,000	$ 1,699	$ 1,495	$ 1,478
$ 20,000	$ 2,660	$ 2,341	$ 2,450
$ 25,000	$ 3,750	$ 3,300	$ 3,558
$ 30,000	$ 4,988	$ 4,389	$ 4,837
$ 40,000	$ 7,958	$ 7,003	$ 7,828
$ 50,000	$11,465	$10,465	$11,335
$ 75,000	$21,685	$20,685	$21,555
$100,000	$32,060	$31,060	$31,930

(d) If your income is $30,000 will your 1975 tax be more or less than your

1974 tax after rebate? _____

Answers to Self-Test

1. Either expansion or shrinkage of the scale gives a distorted and usually false impression of the magnitude (importance) of fluctuations in the curve, visually misleading the reader as to the true significance of the data. (frames 1 and 2)

2. (a) The mean is 50 (550 divided by 11).
   (b) The median is 70 (five numbers above, five below this number).
   (c) The mode (most frequently repeated figure) is 80.
   (frame 3)

3. The statement quoted appears to be based on the assumption that those who saw the film at the private showings and filled out questionnaires represent the majority of citizens. You might have mentioned any of these possible fallacies: (1) the replies to the questionnaire do not necessarily represent the true opinions of those who attended; (2) those who attended were not necessarily representative of those invited; or (3) those invited were not necessarily representative of the majority of citizens. You may have mentioned others. (frame 6)

4. I hope you answered no. Although the blast furnace representing capacity added in the '30's was drawn just two-thirds as tall as the one representing the '40's, the eye sees two furnaces, one of them close to three times

as big as the other. To say "almost one and a half" and to be heard as "three" is what this type of one-dimensional picture often accomplishes. Although only half again as tall, the '40's blast furnace has fattened out horizontally beyond the proportion of its neighbor, and a black bar, suggesting molten iron, has become two and one-half times as long as in the earlier decade. Here was a 50% increase <u>given</u>, but drawn as 150% to give a visual impression of something more like 1500%! (frame 7)

5. Add the numbers to find their total, then divide the total by the quantity of numbers added. This is the same thing as finding the mean, just a difference in terms. (frame 8)

6. (a) the number of individuals (in thousands) employed in Exchange Community
    (b) time, in increments of one year
    (c) 20,000
    (frame 9)

7. right angles (frame 10)

8. Because it is a graphic picture it gives you a quicker overall impression of the <u>trend</u> of the data; it (usually) allows you to read intermediate values not shown in a table; and it shows the <u>range</u> of the values at a glance. (frame 11)

9. Your completed table, with the data from question 10 filled in, should look something like this:

Age	Very Satisfied	Satisfied	Neutral	Dissatisfied	Very Dissatisfied
20-25	12	9	5	2	0
26-30	9	9	4	1	1
31-35	7	7	6	4	3
36-40	6	4	5	4	6
41-45	5	3	4	5	8
46-50	4	2	1	7	11

(frame 12)

10. The data should be filled in as shown above. Several trends are apparent. The number of those "Very Satisfied" is maximum at the youngest age group and decreases consistently to the oldest age group. And, as it would be natural to expect, the number of those "Very Dissatisfied" is minimum (zero, actually) for the youngest age group and <u>increases</u> steadily to a maximum for the oldest age group—the logical inverse of the pattern found for the "Very Satisfied." Also, among the three youngest age groups there is a fairly consistent decrease in numbers from "Very Satisfied" to "Very Dissatisfied"—that is, from left to right. (frame 13)

11. Your curve should be similar to that shown at the right. (frame 14)

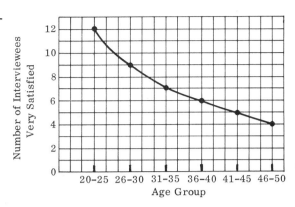

12. (a) overturns, more than double the next highest
    (b) 26%
    (frame 17)

13. (a) 20-24
    (b) 20-24
    (c) 20-24
    (frame 18)

14. curvilinear (frame 19)

15. the area under the curve (frame 20)

16. vertical bar graph or columnar graph (frame 22)

17. column and step curve chart (frame 23)

18. (a) histogram
    (b) pie chart (sectogram or circle graph)
    (c) pictograph
    (frame 24)

19. line graphs, surface graphs, bar graphs, and special graphs (frame 25)

20. (a) 65 grams
    (b) $40^{\circ}C$
    (c) Yes, it certainly appears to be, based on the five points plotted.
    (frame 26)

21. x and y (frame 28)

22. y will increase (frame 29)

23.

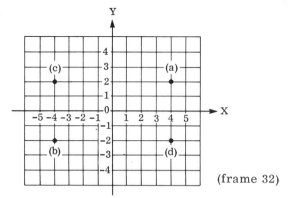

(frame 32)

24. (a)   125 miles
    (b)   5 hours
    (frame 33)

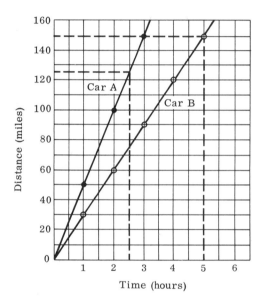

25.

X (frame 35)

26. observed (frame 37)

27. (a) $867
    (b) $1,495
    (c) $2,450
    (d) more
    (frame 39)

CHAPTER TWO
Line Graphs: Straight and Curvilinear

As noted in Chapter 1, the graphs we will discuss in this book are divided into four major categories: line graphs, surface graphs, bar graphs, and special graphs.

Following the general introduction in Chapter 1 we will now examine in more detail the graphs in each of the four major categories. And, as you can see from the title of this chapter, we are going to start with the large family of line graphs.

In this chapter you will learn about the special characteristics, advantages, disadvantages, interpretation, and uses of straight and curvilinear line graphs. When you complete this chapter you will be able to:

- plot and interpret straight line graphs;

- find the slope of, and write the equation for, a straight line;

- recognize and interpret single line, multiple line, and family-of-lines graphs;

- recognize and interpret curvilinear graphs, including simple, cumulative proportion, normal distribution, power, and exponential curves.

STRAIGHT-LINE GRAPHS

1. Before we begin to discuss the various kinds of graphs in greater detail, a word of advice: Any time you feel you are getting lost about where we have been or where we are going, look at the chart in the Appendix at the back of the book. This chart will help orient you. We will discuss each type of graph in the order it appears there.

First we'll review the <u>single-line graph</u>. As you learned in Chapter 1, the single-line graph consists of one straight line representing a linear relationship between two quantities (variables), as shown at the top of the next page. Thus, $y = ax + b$. (We'll discuss this equation in frame 2.)

This graph is the time-speed-distance relationship from frame 26 of Chapter 1. Stated in letter symbols that relationship was, in its basic form, d = s × t (distance equals speed multiplied by time). But remembering from algebra that we can drop the × (times) sign, we condense this to d = st. Thus, a formula, or equation, such as d = 60t (inserting 60 mph as the speed value) is said to be linear because it <u>plots as a straight</u>

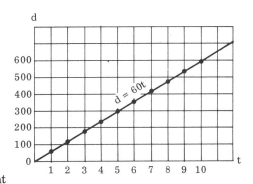

<u>line</u>. If we assume some (constant) speed value such as 60 mph, and then assign various values to t, the resulting values of d correspond directly to the assigned t values. In other words, as t gets larger, so does d. That is, the greater the time, the farther the car will travel at a speed of 60 mph. We found when we tried it that the corresponding values of d and t plotted as a single, straight-line graph. Conversely we can say that a straight-line graph represents a linear relationship between two variables.

As we found in Chapter 1, interpreting a single, straight-line graph is comparatively simple. Reading up vertically from any number value on the bottom (horizontal) scale to the graph and then across, horizontally, to the corresponding value on the vertical scale allows us to find instantaneous solutions of the equation the line represents. Or we can start with values on the vertical scale, go across till we hit the line, then down to the value directly below on the horizontal scale.

Let's look at one more example of this before we go on, just to make sure you have the idea. In physics there is a formula V = IR, where V stands for voltage (in volts), I for current (in amperes), and R for resistance (in ohms). What this formula says is that the voltage in an electrical circuit is equal to the product of the current times the resistance in the wire. If we arbitrarily assume a resistance of 10 ohms in a

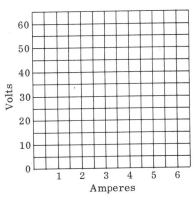

I	V
0	0
1	10
2	___
3	___
4	___
5	___
6	___

circuit, then the formula becomes V = 10(I) (that is, ten times I, the current).

Complete the table of values shown at **the** left of the grid above, substituting the values of current (I) shown, in the formula V = 10(I), to find the corresponding values of V, then plot the pairs of points in the coordinate system shown in the grid. What kind of a graph does this give you?

- -

A straight-line graph, representing a linear relationship between the variables V and I, is the result of the relationship. The completed table of values and plotted points are shown at the right.

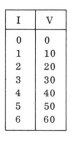

I	V
0	0
1	10
2	20
3	30
4	40
5	50
6	60

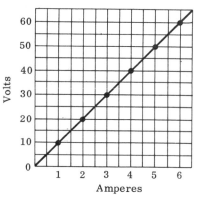

2. In case you are wondering about the formula y = ax + b, which appeared in our definition of a single (straight) line in frame 1, it is the <u>equation of a straight line</u>. This equation is very useful in working with graphs, so let's take time to find out how it was derived and what it means.

The most important characteristic of a straight line is its <u>slope</u>, that is, the tangent of the angle it makes with the horizontal. And this angle in turn is determined by the rate at which y changes with respect to x.

Notice in the figure at the right that for each 2-unit change in x, the value of y changes one unit. Thus, the ratio of change in y to the change in x is 1 to 2, or $\frac{1}{2}$. We say, therefore, that the slope of the line is equal to $\frac{\text{change in y}}{\text{change in x}}$. If we let the letter a represent the slope of a line and the letter b represent its vertical intercept (the point where the line crosses the Y-axis), then we can write the equation for a straight line as y = ax + b.

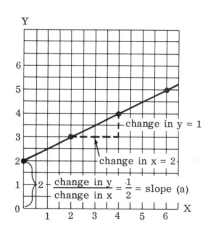

Knowing that, in the graph above, $a = \frac{1}{2}$ and b = 2, write the equation for <u>this</u> straight line. _____

- -

$y = \frac{1}{2}x + 2$. Keep this concept in mind for later use. We will be talking about slope and the straight-line equation later on.

3. We have found so far that the slope of a line is defined as a change of the y value divided by the corresponding change of the x value. Thus, in our familiar example at the top of the next page, $\frac{\text{change in y}}{\text{change in x}} = \frac{120}{2} = 60$ mph.

It is the <u>meaning</u> of the vertical increment (in this case, distance) and of the horizontal increment (in this example, time) that gives the slope its importance as a source of information.

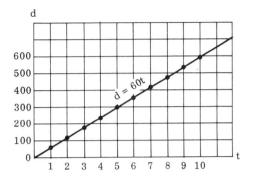

Let's apply the definition of slope to another plot to determine both the slope value and its significance. At the right is a straight-line plot of rectangular area versus length. Study carefully all parts of this graph and see if you can find the meaning of the slope of the line. Ask yourself, What information about this rectangular area does the line slope supply? When you have decided this, then evaluate the

slope numerically. _____

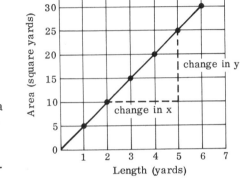

- - - - - - - - - - - - - - - - - - - -

The correct <u>meaning</u> of the slope of the line is that it represents the <u>width</u> of the rectangles. This is so because slope is a ratio of the vertical increment (area) to the horizontal increment (length). And $\frac{\text{area}}{\text{length}}$ = width (here, the slope). To evaluate the slope we take the change in y, 15 square yards (shown by the vertical dashed line), and divide it by the change in x (horizontal dashed line) which is 3 yards, to get $\frac{15}{3}$ = slope = 5 yards = width.

4. A variation of the single-line graph is the <u>multiple-line graph,</u> which is a graph that consists of two or more straight lines representing two or more linear relationships between the same two quantities (variables). We saw an example of this in Chapter 1 (frame 33) where the graphs of two cars, going at different speeds, were compared. The variables—distance and time—were the same for both cars, hence they could be plotted on the same set of coordinate axes.

Examine the multiple line graph at the right. The line sloping upward from 0 represents the motion of a car going east at 50 mph. The line sloping downward from 150 on the vertical scale represents a second car.

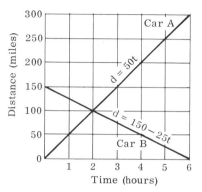

(a) What do you think is the significance of the fact that the line representing car B starts above zero and then slopes downward? _____

_____

_____

(b) Have you any idea (from the graph) when and where the two cars will pass each other (assuming they are traveling on the same road)?

_____

_____

- -

(a) Since the direction of the line is <u>downward</u>, we have an instance of a negative slope (i.e., negative y divided by positive x). And since in the grid sign means direction, a negative slope of 25 mph means that car B is moving in a direction <u>opposite</u>—or negative—to that of car A (i.e., westward), having started at a point 150 miles east of the point from which car A departed.

(b) Where the two lines intersect; that is, two hours after both cars started—100 miles from car A's departure point and 50 miles from car B's departure point.

5. You are most apt to encounter multiple-line graphs in mathematics or in engineering and scientific work. They are not quite as common in statistical work relating to economics, business, or government.

A close relative of the multiple-line graph is the graph known as a <u>family of lines</u>. A family of lines is defined as a graph that consists of a set of straight lines having one characteristic in common. By <u>characteristic</u> we mean that the lines either have the same starting point, the same slope, or the same point of intersection.

Notice the three examples at the right. Each graph contains three lines that share some common characteristic. See if you can identify this common characteristic for each of the three figures.

Figure A _____

_____

_____

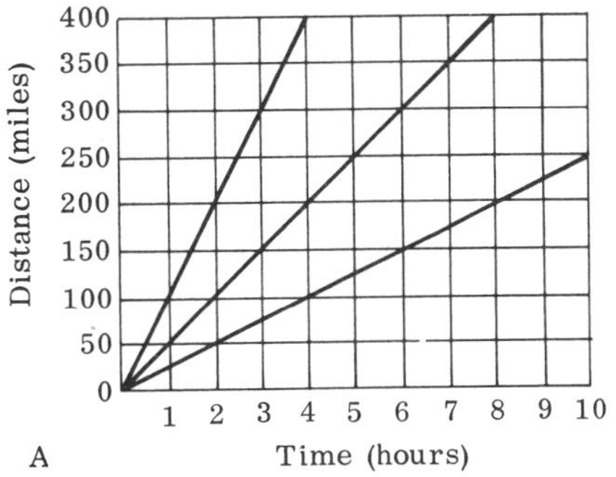

Figure B _____

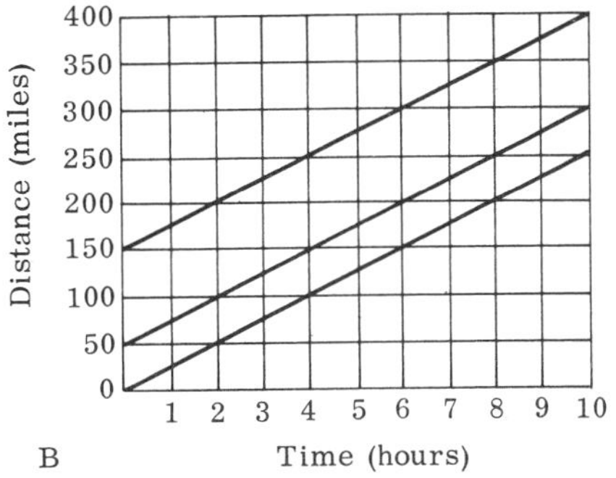

_____

_____

Figure C _____

_____

_____

- - - - - - - - - - - - - - - - -

Figure A: same starting point
Figure B: same slope
Figure C: same point of intersection

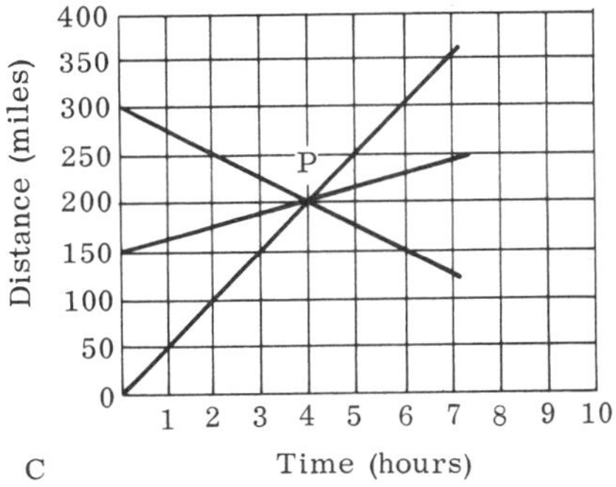

A   Time (hours)

B   Time (hours)

C   Time (hours)

6. A characteristic that remains constant throughout one entire plot (graph of the three lines) but differs from one plot to another is called a parameter. Thus in figure A above each line represents a different speed (a different relationship between time and distance). Each line is constant, thus we know that speed doesn't change because the line direction doesn't change. But the speed does change from line to line (among the three lines). Hence the parameter in this graph is speed, indicated by the differing slope of each line.

Consider figure B above a moment, then answer the following questions.

(a) What is the significance of the fact that the plots are parallel, all having

the same slope? _____

(b) Have you any idea what the parameter is in figure B? State your conclusions here. _____

— — — — — — — — — — — — — — — — — — — —

(a) Each of the three plots represents the same speed.
(b) The parameter is the starting point of each of the three lines, that is, each of them starts at a different point on the vertical axis.

7.   In figure C of frame 5 there are really two parameters—slope and starting point, or vertical intercept—since the slope of each line is different and each starts from a different point on the vertical axis.

Now, what is the value of being able to recognize a <u>family</u> of lines and to distinguish their similar characteristics and parameters? The value lies primarily in the ease with which the line members of the family can be compared. Since only one quantity (usually) varies from one plot to another, comparing the various family members with one another quickly shows the effect of the changing parameter. Thus, in figure A (of frame 5) a check of the distances covered for different times shows the effect of using a <u>different</u> speed for each of the three (imaginary) cars starting at the same point. Figure B on the other hand shows the effect of different starting points for three automobiles having the <u>same</u> speed. Figure C shows the kinds of variations in starting point and slope that can occur when several lines are required to pass through the same point.

Line families also are very useful in graphical solutions of problems, as we saw in frame 2. Suppose, for example, we wanted to have four cars pass the same point at the same time, each car having a different speed and a different starting point. There is a whole family of lines that would satisfy these conditions. See if you can draw in four such lines in the figure at the right.

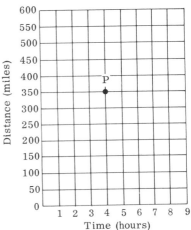

— — — — — — — — — — — — — — — — — — — —

The four lines shown passing through the point P in the figure at the right merely represent <u>any</u> four lines that could be drawn through the point. So long as they have different slopes and different starting points, any four lines would satisfy the requirements. The lines drawn through this point are all of one family that has as its common characteristic the point P, since that point necessarily lies on each of the lines drawn through it.

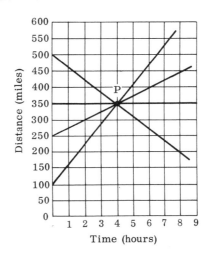

8. At the right is a family of curves developed to show the thickness of various materials required to stop completely beta-ray (electron) radiation. Curves of this type are important in the design of shielding enclosures or surfaces.

For the graph at the right, answer the following questions.

(a) What is the parameter?

_____

_____

(b) What is the nature of the curve shown for each material (i.e., what is each a plot of)? _____

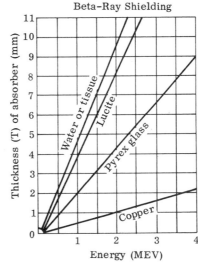

Thickness of typical materials required to stop completely beta-rays of maximum energy

_____

_____

- - - - - - - - - - - - - - - - - - - -

(a) The parameter is simply the material used.
(b) The curve shown for each material is a plot of the thickness of that material required to stop completely beta-ray radiation of different energy levels.

CURVILINEAR GRAPHS

9.   Having looked at some important types of straight-line graphs let us turn our attention now to the curved-line or curvilinear graphs. We have to be careful of our terminology here because mathematicians, scientists, and engineers generally use the term <u>curve</u> to mean either a straight line or a curved line smoothly drawn to connect, in order, a series of plotted points. In statistical work also a "curve" generally is <u>any</u> line that connects a series of plotted points.

   To avoid confusion we will distinguish between straight-line graphs and curvilinear graphs. A straight-line graph is one that consists of one or more straight lines representing a <u>linear</u> relationship between two quantities (variables). A curvilinear graph is a curved line representing a <u>nonlinear</u> relationship between two variables.

   Let us go on then to define the first type of curvilinear graph we will discuss. A <u>simple curve</u> is any one of the class of simple line graphs whose slope is changing continuously at each point on the curve. We worked with a simple curve in frames 35 and 36 of Chapter 1, and in frame 37 we learned to classify such curves as usually representing an observed re-

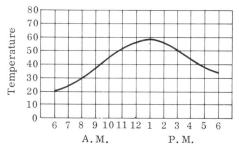

lationship, an empirical relationship, or a theoretical (derived from an equation) relationship. You also have learned that whereas a straight line may (if it starts at the intersection of the axes—i.e., the zero point) represent a constant relationship between the variables, a curved line definitely represents a changing relationship. And since only a straight line has a constant slope, the slope of a curved line is continuously changing.

   With this in mind, and referring to the graph above, answer the following questions.

(a)   What kind of relationship does this curve represent? _____

_____

(b)   Is the relationship between the variables constant or changing? Why?

_____

- -

(a)   an empirical relationship (i.e., a curve fitted to the data by eye)
(b)   changing, because a curve represents a changing relationship between two variables

10. Now we must develop a little statistical background because the next type of curvilinear graph is used most commonly in the field of statistics. As we already have discussed (in Chapter 1) statisticians are concerned with the

problem of analyzing the significance of data derived from many sources. Sometimes these data are obtained by sampling, sometimes by observation, sometimes by testing, and in various other ways. The term sampling can mean a great many things. It may refer to mailing out questionnaires or conducting interviews to get answers to questions concerning people's attitudes or opinions on such topics as politics, sex, the state of the economy, favorite brands of toothpaste, who will be the National Football League champions, or who should be named Woman of the Year. Or it may mean something as mundane as testing every ten-thousandth Whoozit manufactured to ensure that production methods are meeting quality control standards.

A great deal of the information we're concerned with plotting—or interpreting—is derived from various kinds of statistical sampling. For example, one of the things educators and school psychologists are very much interested in is what aptitudes students may have for various vocational careers. To this end they frequently administer aptitude tests and then graph the results. The table below presents the results of one such test.

(1) Score Range	(2) Frequency	(3) Proportion	(4) Cumulative frequency	(5) Cumulative proportion
72 – 74	4	.03	158	1.00
69 – 71	15	.09	154	.97
66 – 68	21	.13	139	.88
63 – 65	21	.13	118	.75
60 – 62	24	.15	97	.61
57 – 59	16	.10	73	.46
54 – 56	23	.15	57	.36
51 – 53	15	.09	34	.22
48 – 50	7	.04	19	.12
45 – 47	3	.02	12	.08
42 – 44	4	.03	9	.06
39 – 41	1	.01	5	.03
36 – 38	3	.02	4	.03
33 – 35	1	.01	1	.01
$\sum$ (sum)	158	1.00		

Now let's find out what this table is all about.

Column 1: These are simply the score brackets, or ranges, of three grade points each, into which the test results have been subdivided for purposes of analysis.

Column 2: "Frequency" refers to the number of students who received a grade within the score range indicated.

Column 3: This column lists the proportion of the whole represented by each figure in the frequency column (that is, the number of students in a score range is compared to the total number taking the test). Thus, looking at the first row of figures, four students received a score in the range of 72 to 74, and 4 divided by 158 (total students taking the test) is .03. Proportion and percentage are essentially the same thing. The difference is that in proportion, the <u>whole</u> is represented by the number 1.00 whereas in percentage the whole is represented by 100. To convert the numbers in the proportion column to percentages we would multiply them by 100 to eliminate the decimal.

<u>Column 4</u>: The cumulative frequency is simply the total of the frequencies in column 2 added to any given point. Notice at the bottom of column 4 that the cumulative frequency is 1. This is because the frequency shown in column 2 is 1. However, the second figure from the bottom in column 4 is 4, the sum of the two lowest figures in column 2. Similarly, the fifth figure from the bottom in column 4 is 12, the sum of 1, 3, 1, 4, and 3—the <u>five</u> lowest numbers in column 2, and so on.

<u>Column 5</u>: The figures in column 5 are also sums—in this case, accumulations of the proportions in column 3. For example, the figure .46 in column 5 is the sum of all the figures in column 3 from the lowest up to and including the figure .10, which is in the same row as .46. (Remember, <u>rows</u> are horizontal; <u>columns</u> are vertical.) The figures in this column have been "rounded off" and so won't always equal exactly the sums of the corresponding figures in column 3.

The summation figures at the bottoms of columns 2 and 3 represent respectively the total number of students tested (test scores) and the total proportion. The Greek capital letter sigma ($\sum$) is used commonly by statisticians to represent sum.

(a)   How many students received a grade of 63–65? _____

(b)   In this same row, how was the figure .13 in column 3 found? _____

_____

(c)   What does the figure 118 in column 4 represent? _____

_____

(d)   What does the figure .75 in column 5 represent? _____

_____

(e)   What is the total cumulative proportion? _____

- -

(a) 21; (b) dividing 21 by 158; (c) total frequency of scores of 65 and below; (d) the sum of the figures in column 3 from the lowest proportion figure up to and including the figure in the 63–65 score range row; (e) 1.00

As pointed out in this frame, if you add the proportion figures in column 3, they don't always total the corresponding cumulative figures in column 5. For example, adding the first three (lowest) figures in column 3 gives us a total of .04. However, the corresponding cumulative total figure (third from the bottom) in column 5 is .03. Don't be alarmed. These occasional slight differences are caused by the rounding off process necessary to making two-digit figures out of proportion numbers that originally were carried to several decimal places.

11. There are a great many numbers in the table shown in frame 10. If you studied them long enough you probably could perceive some general trends. But why wade through these data when it is so easy to plot the significant values and thus have a picture of the trend?

The graph at the right is just such a plot. It was prepared by using the figures in columns 1 and 5 of the table and plotting cumulative proportions against scores. The result is the relatively simple curvilinear relationship you see here. You can see that there has been some curve fitting, such as we discussed in the last chapter (frame 37), in order to develop a smooth curve that would fit the empirical data (that is, test scores versus cumulative proportions). This curve represents the analyst's interpretation of his results.

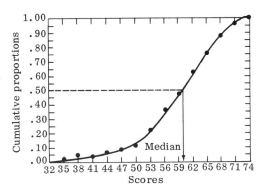

What is the median (middle) score according to the above graph?

_____ How was it found? _____

_____

- -

About 60. It was found graphically by starting at the .50 (middle point) on the cumulative proportions (vertical) scale, moving across horizontally to intersect the curve, then dropping down to the point directly below on the horizontal scale to read the value 60.

12. The curvilinear graph we introduced in the preceding frame is called a cumulative proportion curve. This type of curve is used in statistics to show the cumulative distribution of data in terms either of proportion or percent.

Like the frequency polygon (see Chapter 3, frame 5) and the histogram (see Chapter 5, frame 2), the cumulative proportion curve is another useful way of depicting a frequency distribution. However, having ordinate (vertical axis) values in terms of cumulative proportion or percent, it is particularly convenient for finding the median (50%) and the upper and lower quartiles (25% and 75%). Also note that the graph shown rises most rapidly toward the center of the score distribution and only slightly less rapidly from 68 to 74. The slow acceleration of the graph at the left end of the score distribution (horizontal scale) immediately tells the experienced student that there is a tail (a series of intervals with small frequencies) toward the low end of the score distribution.

Look back at the table in frame 10 (page 69) and see if you can identify each of the points on the graph with the values in columns 1 and 5 from which it was derived. You need to be able to do this if you are going to be able to read and interpret tables and graphs of this kind. Let's try a couple.

There are fourteen rows of data in our table (on page 69) and fourteen corresponding data points on the graph. The horizontal scale of the graph consists simply of the range of scores. To plot the first point, then, we first locate the lowest cumulative proportion figure, .01, on the vertical scale (as nearly as we can estimate its position). Using a straight-edge to guide us we then move across, horizontally, to a position that is directly above the number 35 on the Scores scale. The value 35 indicates, of course, the lowest score range in column 1. (We will use the upper values in each case of the three-point spread of values in column 1 in order to be consistent. Using the lowest values or middle values would be equally correct; it would only result in shifting the curve slightly to the left but would not alter its shape.)

The dot you find located in this position in the lower left corner of the chart represents the first point of the curve. Using each of the remaining 13 sets of score and cumulative proportion values from the table, locate the corresponding points on the graph, numbering them in sequence from 2 through 14.

- -

Your numbered points should be as shown at the right.

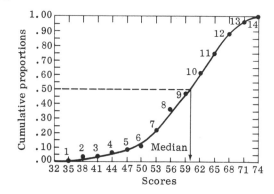

13. Another curve from statistics—one you doubtless have seen many times—is the so-called <u>normal distribution curve</u>. This is a bell-shaped curve, symmetrical about the mean, representing the distribution of sampling data. An example is shown at the right.

As we mentioned earlier, statisticians are very much concerned with the matter of sampling and sample size. If we wanted to know how many people in the United States have blue eyes, for example,

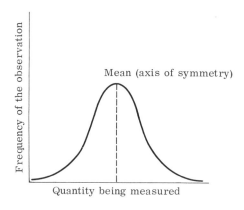

the only way we could obtain an absolutely accurate count would be to actually observe all the many millions of individuals involved. Since this is obviously impossible, we would try instead to select a "representative" or "random" sample of the population for observation. (How to avoid choosing a <u>biased</u> sample is a subject in itself and one beyond the scope and intent of this book. See any good textbook on statistics for this kind of information in case you are interested.)

We might, for example, select a group of 100 individuals in a certain part of the country and discover that 67 out of the 100—that is, 67%—had blue eyes. Could we accurately conclude from this that 67% of the total population has blue eyes? Not if our sample came from northern Minnesota or some such area where the population is of predominately Scandinavian descent.

To avoid this obvious kind of bias let's assume we select instead groups of 20 individuals each from 50 different locations (the 50 states, perhaps) throughout the United States. This would not only increase our sample size to 1000 but would also provide a much broader cross section of ethnic types. Our data might look something like that at the right.

Plot the pairs of values in the table on the grid system below and draw as smooth a curve as you can through the points. (Note: Some of the points will <u>not</u> lie on the curve.)

Number with Blue Eyes in Each Group	Frequency of Observation*
less than 5	0
6	1
7	1
8	2
9	4
10	6
11	7
12	8
13	7
14	6
15	4
16	2
17	1
18	1
over 18	0
Total 50	

*Number of groups in which observed.

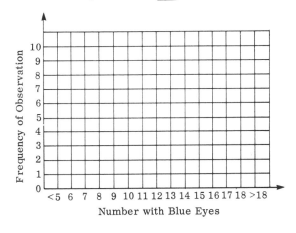

- -

Your curve should look about like the one shown here. The points at 6, 12, and 18 will not fall on a smooth curve <u>that best fits the other points.</u>

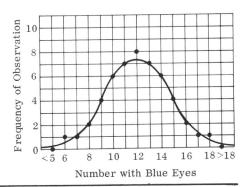

14. The word <u>normal</u> as it is used in connection with a curve such as the one in frame 13 does not mean the opposite of abnormal. Although its precise meaning as used in statistics is somewhat more complex, in simplest terms it refers to the fact that the <u>area under the curve is distributed symmetrically about the mean.</u>

The tendency for sampling distributions to take on a common shape as larger samples are considered is a useful and important fact in statistical work. In fact there is a general law that, almost regardless of the shape of the original population, the shape of the sampling distributions derived from it will be approximately normal. This law can be proved mathematically (which we are <u>not</u> going to do) and is known as the <u>central limit theorem.</u>

However, this fact of normal distribution of samples results in enormous simplifications. It means that a great many important problems can be solved by this single pattern of sampling variability. A wide class of phenomena turn out to be roughly normally distributed if a frequency distribution is made from a large enough number of cases—for example, the weights and dimensions of plants or animals of a given species, or of objects produced under similar conditions; the number of words on printed pages of a given dimension and type size; intelligence quotients; and baseball batting or fielding averages.

The characteristic shape of a normal distribution curve, as we have seen, is a humping up in the middle, falling off in either direction, at first with increasing steepness and then with decreasing steepness.

On the other hand many kinds of data depart widely from normal. The incomes of families or individuals, the number of cars per family, or the number of wheels per car, are examples. The fundamental, statistically important fact remains, however, that the measures computed from samples usually tend to be normally distributed whether or not the original data are normally distributed. Thus, the larger the sample the more likely it is that most of the cases will cluster around the average.

This means, for example, that if the average height of an American man is 5 ft. 8 in., then 7 out of 10 men will not be more than 3 in. taller or shorter than this. It is this fact that makes possible the production of ready-made clothing. All physical characteristics of people follow this pattern.

There are several significant statistical terms and concepts necessary to use tables of normal distribution intelligently, but this is not the time or place to go into them. Therefore, if you come across a normal curve in connection with your work and want to work with it, you are urged to consult the Wiley Self-Teaching Guide <u>Statistics</u> or some equally good reference book for help.

Which of the curves shown at the top of the next page would you say represents a normal distribution? _____

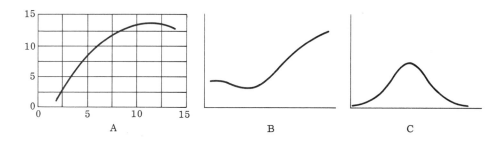

A   B   C

- -

curve C

15. So far in our exploration of curvilinear graphs we have looked at simple curves, the cumulative proportion curve, and the normal distribution curve. Now we are going to consider two additional types of curve: the power curve and the exponential curve.

These curves are both mathematically oriented, in the sense that they are derived from mathematical equations. So if you are not mathematically oriented, you may wish to skip the rest of this chapter and go directly to Chapter 3. But if you are interested in at least knowing what these curves look like, how they are defined and used, then you are urged to read at least the portion of each frame dealing with these aspects of the curves and skip over the mathematical details given.

Let's move on, then, to the first of these curves. The power curve is a curve in which one coordinate (dependent variable, y) is related to some power of the other coordinate (independent variable, x). Thus, $y = ax^b$.

The relationship between two variables in many aspects of the physical sciences, as well as psychological research problems and problems in economics, is represented by this type of curve. For example, the curve shown here actually is one-half of a curve known as the parabola (which we will discuss a little later). And it has such useful properties as describing the path through the air of an object thrown or fired, the shape of the supporting cables of a suspension bridge, or the shape of the reflectors used in automobile headlights.

Although the significance of the relationship between the two variables in a power curve will depend upon the nature of the variables themselves, there is one observable element in that relationship that is totally unrelated to the nature of those variables, and that is the shape of the curve. Thus, the greater the value of b in the equation $y = ax^b$, the steeper the curve.

The curve shown at the right is the graph of $y = x^3$. Notice the increase

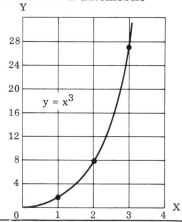

in the steepness of this curve compared with the plot of $y = x^2$ shown in the following graph. We can't even plot the value of y for x = 4 because it falls off the chart. And when x = 3, y = 27, whereas in the graph below when x = 3, y = 9. Your eye alone, then, will tell you something about the value of b in the curve's equation, $y = ax^b$.

The second example, shown at the right, is the graph of the equation $y = x^2$. Comparing this equation with the general equation $y = ax^b$ shows us that, in this case, a = 1 and b = 2. (a and b are arbitrary constants that can represent <u>any</u> number. Don't let the fact that we used a to represent the slope of a straight line in frame 2 lead you to think it means that here.) Putting the equation $y = x^2$ into words we would read it as "y equals x to the second power," hence the name <u>power</u> curve.

Use this graph to find the approximate value of y when x equals 2.5.

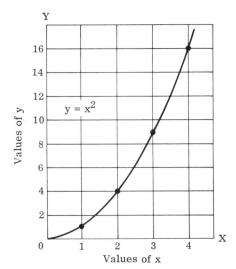

----

about 6.25

16. It's not too easy to read values on a graph of this kind, is it? Particularly at the low and high ends. Curves that move steeply upward have a built-in readability problem, regardless of how large they are drawn. However, an interesting thing happens if we take the log (logarithm)* of our general power curve equation: $y = ax^b$ becomes log y = log a + b log x.

What's so interesting about this? The interesting thing is that in this form it is the equation of a straight line. (Remember from frame 2 that y = ax + b, or y = b + ax, which is the same thing.) Let's place one equation above the other so you can see the relationship more clearly.

   (1)   y = b + ax

   (2)   log y = log a + b log x

Putting them each into words may help.

   (1)   The value of the dependent variable equals the sum of a constant plus the product of the independent variable and another constant.

(2)   The value of the log of the dependent variable is equal to the log of a constant plus the product of the log of the independent variable and another constant.

Now, of what use is this relationship? Can you think of any possible useful implication in this fact? If so, state it below.

_____

_____

_____

- -

The useful implication is that if, instead of plotting the curvilinear relationship $y = ax^b$, we plot the logarithmic relationship $\log y = \log a + b \log x$, the resulting graph will be a straight line, or linear relationship between the variables and, therefore, much easier to read accurately. However, don't be too concerned if you had no notion whatever as to what the answer should be. Only the reader with a fairly recent background in math would be expected to observe this fact.

17. Let's apply this approach so you can see for yourself what happens. For our example we'll use the specific power curve $y = x^2$, the same curve we plotted in frame 15. However, this time we're going to plot the <u>log</u> of $y = x^2$. Comparing $y = x^2$ with the general expression $y = ax^b$, you can see that we have let $a = 1$ and $b = 2$. Taking the log of both sides of our equation, therefore, we get $\log y = 2 \log x$.

The table at the right contains all the values (to the nearest two decimal places, taken from the log tables in the Appendix) we need in order to be able to plot both graphs, that is, the curvilinear graph $y = x^2$ in frame 15 and the logarithmic version of it. This will enable us to compare the two.

(1)	(2)	(3)	(4)	(5)
x	log x	2 log x	y	log y
1	0	0	1	0
2	.30	.60	4	.60
3	.48	.96	9	.96
4	.60	1.20	16	1.20
5	.70	1.40	25	1.40
6	.78	1.56	36	1.56
7	.85	1.70	49	1.70
8	.90	1.80	64	1.80
9	.95	1.90	81	1.90
10	1.00	2.00	100	2.00

Columns 1 and 4 contain the values used to plot the graph of $y = x^2$ in frame 15. Use the values in column 3 and 5 to plot the graph of $\log y = 2 \log x$ on the coordinate system at the top of the next page.

What kind of graph
does this give you?

_____

_____

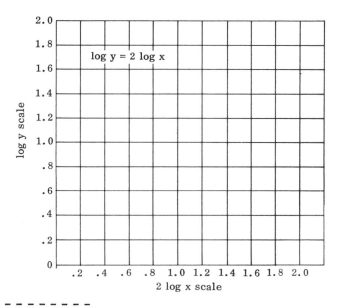

— —

Your graph should look
generally like this one.
As you can see, it gives
you a straight-line plot,
thus confirming our
prediction.

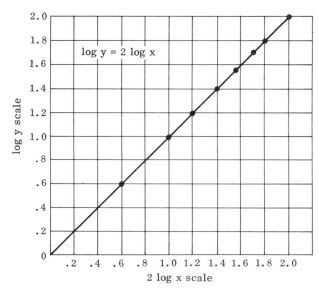

Scales such as those used to plot the logarithmic form of the power curve
$y = x^2$ would, of course, be rather awkward to work with for most plotting because we would have to keep looking up logarithms of the x and y values in order to be able to plot the equations. For this reason a special kind of chart
(logarithmic) has been developed to eliminate this problem. However, we will
not use these special graph papers in this book.

18. Our second curvilinear graph is a close relative of the power curve. It is the exponential curve, a curve defined by an equation in which the independent variable appears as an exponent. Whereas in the power curve the independent variable (x) is the base (as in the equation $y = x^2$), in the exponential curve it is the exponent (as in $y = 10^x$). As with the power curve, you will sometimes find the trend of plotted points best represented by a special curve—the exponential curve, in this case. This can occur in the field of statistics related to psychological research, economics, consumer trends, or the physical sciences, for example. In the exponential curve we would have the equation $y = ae^{bx}$ (in natural logarithms) or $y = a(10)^{bx}$ (in common logs). (The value of e is 2.718.) Here, we will confine ourselves to the equation $y = a(10)^{bx}$.

If we let a and b both equal 1 in the above equation, we get the exponential curve shown at the right when we plot it. Here again we have a curve that is difficult to read with any great accuracy at the low and high ends. As in the case of the power curve it would be most convenient if we could find some way to straighten it out mathematically. Fortunately, as with the power curve, there is a way.

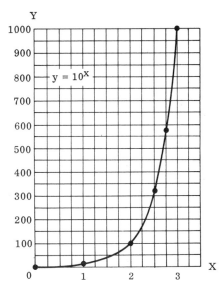

Taking the logs of both sides of the formula $y = a(10)^{bx}$, we get log y = log a + bx. (The number 10 drops out because the log of 10 to the base 10 is 1.) Now if we again let a and b both equal 1, our new formula becomes log y = x. (The log a term drops out because the log of 1 is zero.)

All this last formula says is that the log of y is numerically equal to the value of x. That is, if x is 2, log y also is 2. If x is 5, log y is 5, and so on. To plot the relationship log y = x, then, we don't even need to go to the trouble of preparing a table of values, for whatever x is, log y is. So with this simple approach in mind, use the coordinate system at the right to plot log y = x. What does the resulting graph look like?

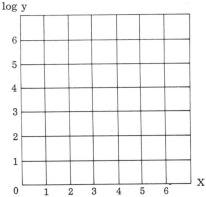

- - - - - - - - - - - - - - - - - - - -

You should have a straight-line graph starting at zero and moving upward at a 45° angle from the horizontal axis. (The fact that the x and log y values are the same and the line is therefore half-way between the vertical and horizontal axis automatically means its slope will be 1.)

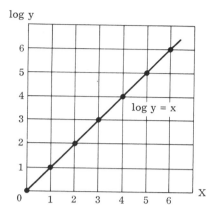

SUMMARY OF CURVILINEAR GRAPHS

Since we have covered eight different types of curvilinear graphs, plus several variations of those types, a brief summary should help refresh your recollection of just where we have been during the past 30 or so frames. Briefly, these are the curves we have examined, in various degrees of detail, depending upon their complexity and potential usefulness.

Simple Curve: Any of the class of simple curves in which the ratio of the variables is not constant.   (frame 7)

Cumulative Proportion-Percentage: A curve used in statistics to show the cumulative distribution of data in terms either of proportion or percent.
                                                                    (frames 8-10)

Normal Distribution: A symmetrical, bell-shaped curve representing the distribution of sampling data.   (frames 11, 12)

Power Curve: A curve in which one coordinate (dependent variable, y) is related to some power of the other coordinate (independent variable, x). Thus, $y = ax^b$.   (frames 13-15)

Exponential Curve: A curve defined by an equation in which the independent variable appears as an exponent. Thus, $y = ae^{bx}$ (in natural logarithms) or $y = a(10)^{bx}$ (in common logs).   (frame 16)

Before proceeding to the Self-Test that follows you would do well to go back and review any of the straight and curvilinear graphs that still are not clear to you. You should not expect to have learned all there is to know about them after this brief introduction. But the name of the graph should at least call to your mind its general appearance and chief characteristics.

SELF-TEST

1. A straight-line graph represents what kind of relationship between the two variables? _____

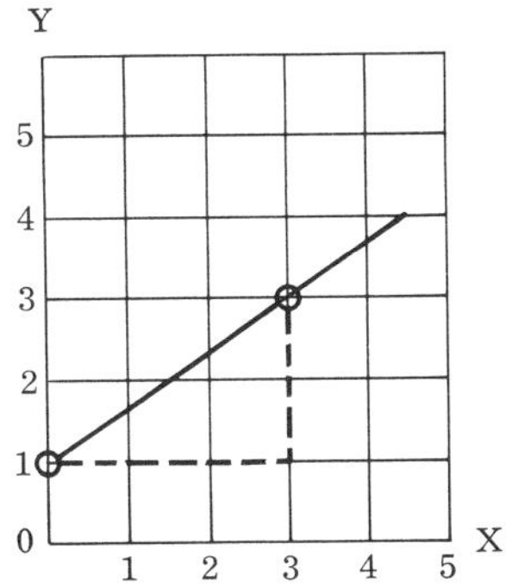

2. In the figure at the right:

(a) What is the change in y? _____

(b) What is the change in x? _____

(c) What is the slope a (i.e., ratio of y to x)?

_____

(d) What is the vertical intercept b? _____

(e) What is the equation of <u>this</u> line, based on the general equation y = ax + b?

_____

3.   In the figure at the right:

(a) What does this graph repre-sent? _____

(b) What is the slope of line 1?

_____

(c) If the building code required a minimum pitch (slope) of 2:3, could you make it by fol-lowing the slope shown in line 2? _____

_____

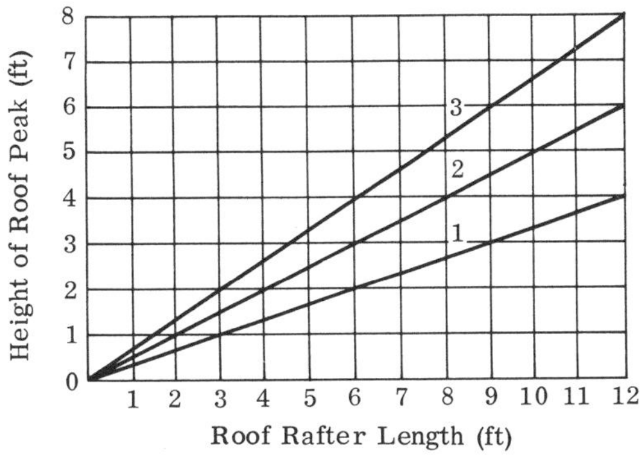

4.   What characteristic does the family of lines shown in problem 3 have in common? _____

5. What characteristic do the three curves (lines) in the figure at the right have in common?

_____

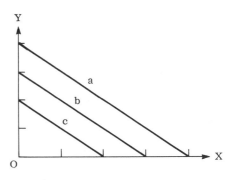

6. (a) In the chart below, what is the parameter of the family of lines?

_____

(b) This chart is intended to show respiratory water loss (by astronauts) while working under stress. Can you tell what variable factors the

lines represent? _____

_____

_____

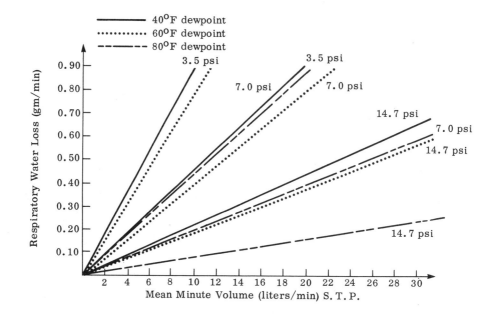

7.   What kind of relationship does the curve at the right represent (observed, empirical, or theoretical)?

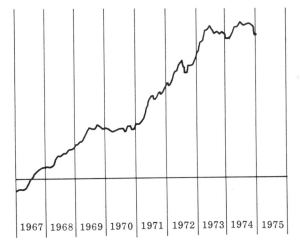

_____

8.   In the cumulative proportion graph below:

(a)   What does the vertical scale represent? _____

(b)   What is the median rating for vanilla ice cream? _____

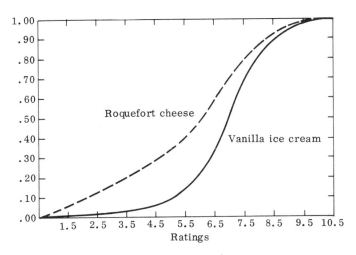

9.   Answer these questions about the curve at the right.

(a)   Identify the curve.

_____

(b)   What does it represent?

_____

(c)   Does it usually assume this shape?

_____

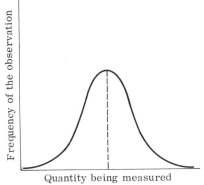

10. In the power curve at the right, what is the approximate value of y when x = 2.5? _____

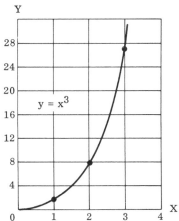

11. The logarithmic version of the equation $y = x^2$ is log y = 2 log x. (True or False) _____

12. The power curve will plot as a straight line on a coordinate system where (one scale/both scales) _____ are logarithmic.

13. Answer the following questions about the curve shown at the right.

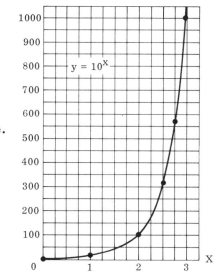

(a) The curve is known as an

_____

(b) On a coordinate system where the vertical scale is logarithmic, this curve plots as a (straight/curved)

_____ line.

(c) What is the value of y when x = 2.5?

_____

Answers to Self-Test

1. linear (frame 1)

2. (a) 2; (b) 3; (c) $\frac{2}{3}$; (d) 1; (e) $y = \frac{2}{3}x + 1$
    (frame 2)

3. (a) Three different slopes of roof pitch ($\frac{\text{vertical rise}}{\text{horizontal run}}$).

    (b)   4:12 or 1:3 ($\frac{1}{3}$)

    (c) No, because this is only 1:2; you would have to use slope 3.
    (frames 3, 4)

4. same starting point (frame 5)

5. They have the same slope. (frame 6)

6. (a) a common starting point
    (b) different dewpoints and atmospheric pressures (psi)
    (frame 6)

7. observed (frame 9)

8. (a) Proportion (cumulative)
    (b) about 7.0
    (frame 11)

9. (a) normal, or bell-shaped, curve
    (b) the distribution of sampling data
    (c) yes
    (frame 13)

10. about 16 (frame 15)

11. True (frame 17)

12. both scales (frame 17)

13. (a) exponential curve
    (b) straight
    (c) about 320
    (frame 18)

CHAPTER THREE
Line Graphs: Zigzag, Step, and Special Scale

In the last chapter we looked at two major types of line graph, namely, straight and curvilinear. Continuing our investigation of line graphs we come to the three additional categories: zigzag, step, and special scale. Just as with the graphs we saw in Chapter 2, many of those you will see here will look familiar to you because you have seen them used in financial reports, Government statistics, technical publications, or similar sources.

What we want to do now, however, is examine them in enough depth so that the <u>next</u> time you see one of them you will be able to categorize it mentally and remember its special properties and purposes. This chapter will teach you how to identify, interpret and plot the following:

- zigzag graphs, including simple zigzag, multiple zigzag, cumulative, cumulative deviation, and frequency polygon;

- step graphs, including simple step and multiple step;

- special scale graphs, including repeat time, multiple time, multiple amount, supplemental amount, and index scale.

ZIGZAG GRAPHS

1. A very informal title, the term <u>zigzag graph</u> is descriptive and should help you to distinguish it from the other graphs. It is also commonly known as a slope-curve chart, but this title may be confusing to you because it suggests the slope we discussed in connection with straight-line graphs. Hopefully, therefore, you will agree to accept the name zigzag.

A <u>simple zigzag curve</u> is made by drawing a line directly from each plotted point to the next. This curve suggests that changes from point to point are gradual or continuous and

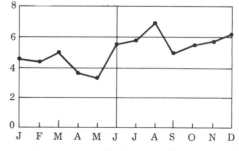

is usually the best way to show data that have "carryover" from one time to the next. This type of time series is sometimes called "as of" data because it measures how things stand at points in time.

In business, zigzag charts typically are used to show month-end inventories, strength, unliquidated obligations, year-end prices or yields, and similar data. However, when fewer than four or five time points are plotted, a column chart usually is better. They can be used equally well to present changes in some variable with time in scientific and engineering work. As with all the curves we will discuss, zigzag curves are used in many fields.

The curve at the right, for example, shows the percentage of test subjects reporting symptoms of acute altitude sickness (hypoxic symptoms) during each of 20 days of continuous exposure to air at an altitude of 12,500 ft. An abrupt and definite indication of change is observable for each of the days charted. A zigzag graph will not, of course, show abrupt changes unless they exist inherently in the data. But when the changes are there they show up clearly.

Would you say, from the graph at the right, that subjects were tending to adjust to the high altitude environmental conditions with time, or not? How did you arrive

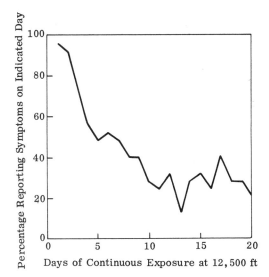

at your conclusion? _____

_____

- - - - - - - - - - - - - - - - - - - -

Generally speaking they seem to be adjusting since the number reporting adverse symptoms has fallen from an initial high of about 98% to an average of about 30% during the last 10 days.

2. A first cousin of the simple zigzag curve is the <u>multiple zigzag curve</u>. It is a graph that brings together two or more related, simple zigzag curves.

The curves may be interdependent, such as a total and its components, or independent, such as two or more totals. They can even be nondependent, such as an actual result compared with a forecast or estimate.

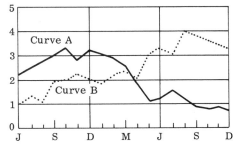

Whatever its purpose this type of chart should not try to compare too many things at once. A chart with four curves is not necessarily twice as useful as one with two curves! In fact, it may be less useful, especially if the curves criss-cross or run together.

Where there are many curves, or several curves of the same magnitude, it is often better to break a single chart into two or more charts with the same scales. When graph makers are not that thoughtful, we sometimes must examine a graph very closely to distinguish the curves from each other.

At the right is a multiple zigzag graph composed of four related quantities: Dow Jones averages in current dollars, Dow Jones averages in constant dollars, yield in constant dollars, and profit on stockholders equity. Despite

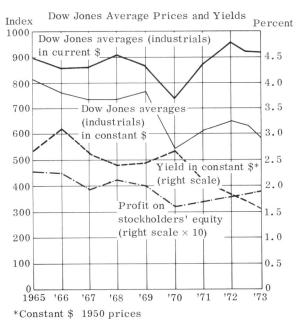

Dow Jones Average Prices and Yields

*Constant $ 1950 prices

the fact that there are four curves here, differentiation (and hopefully clarity) between them has been maintained by the use of a different type of line (heavy, light, dashed, dotted) for each curve and by the written legends attached to each curve.

Notice that this is also an example of a double scale graph. The horizontal scale is, of course, a time scale in years. However, there are two vertical scales. The scale along the left edge of the graph is an index scale (that is, an arbitrary scale, not an absolute scale). The scale along the right edge is a percentage scale. Values of the two top zigzag curves are found on the left vertical scale and values of the two lower zigzag curves on the right scale. This is a convenient technique for comparing the movements of related factors that are measured in different units. Thus the four curves are:

(1) Dow Jones industrial averages in current dollars (read on left scale).
(2) Dow Jones industrial averages in constant (i.e., 1958 prices) dollars (read on left scale).
(3) Yield in constant dollars (percent; read on right scale).
(4) Profit on stockholders' equity (percent times 10; read on right scale).

Use this Dow Jones graph to answer the questions which follow.

(a) What was the approximate value of the Dow Jones industrial averages in current dollars for the year ending in 1968? _____

(b) What was the profit on stockholders' equity at year ending 1969?

_____

(c) What was the Dow Jones industrial averages index value in constant dollars at year ending 1973? _____

(d) What was the yield in constant dollars at year ending 1966? _____

– –

(a) about 900; (b) about 2%; (c) about 600; (d) about 3.1%

3.   We have looked at zigzag charts that showed the status of some quantity at a point in time—at the end of a month, quarter, year, and so forth. Thus, the top curve in the last graph showed the value of the Dow Jones industrial averages in current dollars at the end of each year, from 1965 to 1973.

The next kind of zigzag graph is designed to show cumulative values. The <u>cumulative curve</u> is a chart that presents a running total, that is, each point shows the total to date.

Each point on a cumulative curve shows the aggregate to date— the total for the current period plus all earlier periods. Because it

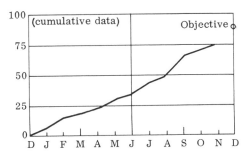

focuses on the cumulative picture, rather than on the amounts for each period, the cumulative chart is an excellent way to show progress toward an annual objective or goal such as the collection or expenditure of funds. Similarly, <u>multiple cumulative charts</u> are effective for comparing results "so far this year" with those for the same period last year, for example.

The process of adding the results for each successive period tends to smooth a cumulative zigzag curve. It is used mostly where the cumulative total is likely to stay the same or increase, but not decrease. If it is not obvious from the nature of the subject that cumulative data are required, the chart should be labeled, "Cumulative from _____," or "Cumulative Data," as shown in the second figure on the next page.

The graph at the top of the next page is a dual cumulative curve showing the aggregate contributions over a period of 12 years of members of an investment club (lower line) and the net asset value of the investment portfolio for the same period. Contributions were made on a regular basis and the accumulated funds invested promptly. (The dip in net contributions in 1959 and 1960 occurred when members declared themselves a dividend of $900 each.)

What did the accumulated contributions total and what was the net asset value of the portfolio as of 1966?

_____

_____

- - - - - - - - - - - - - -

approximately $9,000 and $25,000 respectively

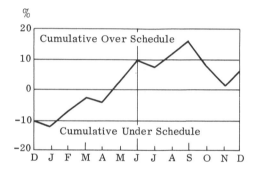

4. Another variation of the cumulative curve is the <u>cumulative deviation curve chart</u>. This is a chart that shows cumulative differences, or deviations, from a reference or base point such as a schedule or a budget.

Like any cumulative chart this one shows the "total to date" at any point on the curve. This type, however, measures cumulative <u>differences</u> or deviations—for example, net gain or loss—from budget or allowance. Hence this kind of curve is likely to go down as well as up, because the net change may be a decrease or an increase.

Notice in the schedule chart shown above that the zero-reference line appears as the horizontal line in the middle of the chart, with positive vertical scale values above and negative values below this line. The scale range in this case is from minus 20% to plus 20%.

From the chart, how much would you say the:

(a) Cumulative over schedule was at November? _____

(b) Cumulative under schedule was at March? _____

(c) At what month was the largest cumulative over schedule percentage reached? _____

(d) At what month was the largest cumulative under schedule percentage reached? _____

- -

(a) about 2%; (b) about 3%; (c) September; (d) January

5.   The last type of zigzag graph we are going to discuss is the <u>frequency polygon</u>. What is it? Let's see. The frequency polygon is a curve used in

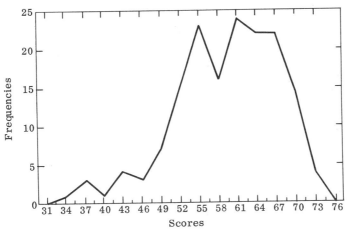

statistics to display the frequency distribution of data samples. The shape tends to approximate that of a normal curve as the sample size increases.

In order to talk about this curve we will need to refer back to a concept we covered earlier, namely, analyzing the significance of data derived from sampling. So turn to Chapter 2, frame 10 (page 69), and look again at the table we discussed there.

You will recall that we used the information contained in columns 1 and 5 to develop the cumulative proportion-percentage curve presented in frame 11 (Chapter 2). However, something the statistician is even more anxious to get a look at is a graphic portrayal of the frequency distribution of his data. To get this we simply plot the mean (middle in this case) score range values from column 1 against the frequency values shown in column 2 to obtain the graph that appears above.

For which middle score range value is the dip in the curve (valley) greatest? _____

- - - - - - - - - - - - - - - - - - - -

58

In case your recollection of geometry is dim, a polygon is a <u>closed</u> figure bounded by straight line segments as sides. The figure in frame 5 doesn't <u>look</u> much like the familiar polygons such as the square, the parallelogram, the pentagon, or the hexagon. But it does, nevertheless, qualify technically as a polygon if we acknowledge that <u>the horizontal base line of the graph forms the otherwise missing side</u> to close the figure.

6.   We have discussed five types of zigzag chart. See if you can identify correctly each of the examples shown at the top of the next page.

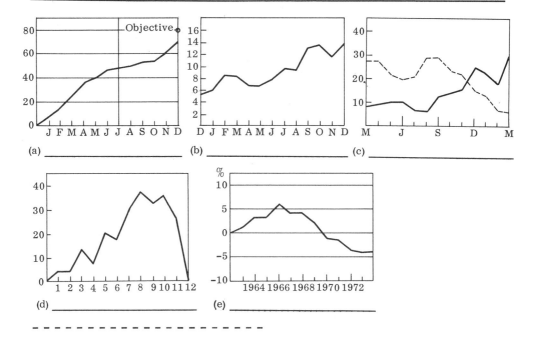

(a) _____ (b) _____ (c) _____

(d) _____ (e) _____

- - - - - - - - - - - - - - - - - - - -

(a) cumulative—shows running total to date at each point; (b) simple zigzag—shows changes from point to point; (c) multiple zigzag—compares two or more zigzag curves; (d) frequency polygon—shows frequency distribution of data; (e) cumulative deviation—shows cumulative deviations from schedule

STEP GRAPHS

7. Now let's examine a simple step curve. This is a graph made by drawing a horizontal line through each point and connecting the ends of these lines by vertical lines. Its purpose is to show averages or other measures that apply over periods of time or for discrete groups of people or products. It is similar to the histogram (see Chapter 5) with the vertical lines removed. It is 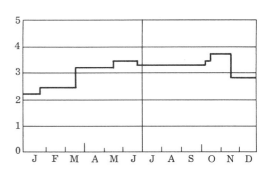 especially good for data that change abruptly at irregular intervals, such as allocation of funds or population data. Step curves are better than zigzag curves for showing "period" data and often are used for this purpose, especially when the time series is a long one. Zigzag charts imply a linear change between two points while the step curve implies an abrupt change between time periods.

The graph at the right is an example of a step curve used to show the population in the United States and in the Armed Forces abroad in July 1962.

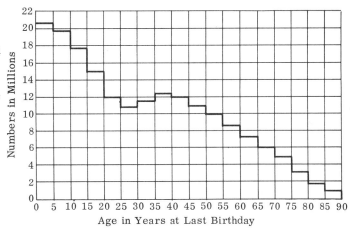

Age in Years at Last Birthday

According to this chart, what was the smallest age group of those between 20 and 45 years of age? _____

- - - - - - - - - - - - - - - - - - - -

the group in the 25–30 age range

8. A logical extension of the simple step curve is the multiple step curve. This is a graph made up of two or more simple step curves. It serves the same general purpose as the multiple zigzag curve but is harder to follow if it contains more than two curves.

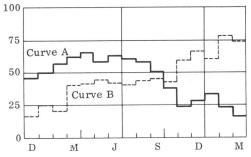

Multiple step curves are difficult to read if they cross back and forth several times. Two or more step curves can be shown on the same chart, however, if they do not overlap or if they cross cleanly, such as those shown above. A multiple zigzag curve is easiest to read if the different curves are coded (e.g., solid line, dashed line, dotted line) as well as identified by name.

The graph at the right shows the budgets of three different departments of the government over a period of ten years.

See if you can answer the questions at the top of the next page which pertain to this graph.

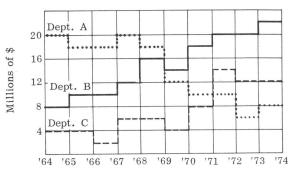

(a) Which department consistently hád the highest budget? _____

(b) Did Department C's budget show a net increase or decrease over the ten-
     year period? _____

(c) Which department had the highest budget during 1968? _____

(d) Which department had the lowest budget during 1972? _____

- -

(a) None throughout the 10-year period. Department A was highest until 1969,
and department B's budget was highest thereafter. (b) increase; (c) Department A; (d) Department A

SPECIAL SCALE GRAPHS

9. Some other types of graph require a special kind of scale or scale treatment. Most of these special scale graphs vary widely in purpose and use.
Because these graphs assume some background from the user, they are best
used in specific analytical situations.

Taken together, this group of
charts demonstrates how effectively a particular type of chart can
focus on the key facts to be com-
municated to a reader.

A repeated time-scale chart
is a graphic method of showing a
long curve as a series of short
curves, such as monthly data over
a period of several years.

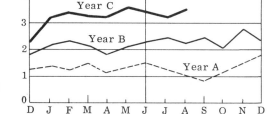

As shown in the chart, month-
ly data for a period of several years can be shown as several one-year pieces
plotted on a one year time scale. This method also permits easy comparison
of comparable time points—of results "so far this year" with results for the
same period last year or some earlier year. It is especially useful for comparing monthly data that have a seasonal pattern (generally high at certain
times of the year and low at others), such as retail sales, inventories, or em-
ployment.

In the graph above, the two lower curves (the dashed line and the lighter
solid line) represent one full year of data each. However the top curve (Year
C) portrays results for the year to date through the month of August. From
the period January through August, then, it is possible to compare operating
results for three years; and from September through December, operating results for two years.

Although this chart appears somewhat similar to the multiple zigzag
graph, they differ in this respect: Whereas in the multiple zigzag the plotted
curves represent different quantities (although related in some way) for

comparison, the repeated time-scale chart represents only <u>one</u> quantity (variable), broken into annual segments for comparison purposes.

Assuming the vertical scale represents monthly sales in millions of dollars, what was the sales volume for June in:

(a) Year A? _____

(b) Year B? _____

(c) Year C? _____

- - - - - - - - - - - - - - - - - - - -

(a) approximately 1.5 million dollars; (b) approximately 2.3 million dollars;
(c) approximately 3.4 million dollars

10. Another special scale graph is the <u>multiple time-scale chart</u>. It is similar to the repeated time-scale graph except that it compares <u>nonrepeating</u> time periods.

As shown in the figure at the right, this kind of presentation is used to compare results or contradictions during two similar historical periods. Although it is an extremely valuable chart when used

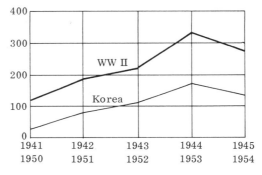

properly, this type is much more difficult to design than a repeated time-scale chart. The main problem is in deciding how to match the periods. For example, in comparing World War I and World War II, the time scales could begin at 1917 and 1941, 1914 and 1940, or 1914 and 1939. Depending on the subject, any of these could be used.

The chart above depicts a comparison between World War II and the Korean war. If we were to assume that the vertical scale represents thousands of U.S. troops engaged in active combat, how many <u>more</u> troops were in active

combat in 1944 (WW II) than in 1953 (Korea)? _____

- - - - - - - - - - - - - - - - - - - -

approximately 150,000 more (330,000 minus 180,000)

11. The <u>multiple amount-scale chart</u> is a graph used to compare two or more curves that are measured in different units, or are measured in the same units but are too far apart in size for easy comparison.

More than two amount scales should not be used with this type of chart. If the aim

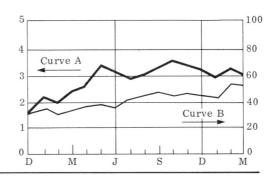

is to compare <u>change</u> or <u>growth</u>, the zero line of both scales should coincide (either on or off the chart), and the scale intervals should be selected in such a way that both curves meet. This provides a common point of departure and makes comparison more meaningful. Often this can best be accomplished by converting both series to a common base so that only one scale is necessary.

If the purpose of the chart is to compare the <u>pattern</u> of two or more curves—without regard for their size—then the scales need not coincide at zero.

Notice in the graph at the beginning of this frame that the left and right vertical scales are different, although they have a common zero point. Curve A takes its values from the left scale, and Curve B from the right scale. Curve B, for example, might represent company sales and Curve A profits.

Based on these observations would you say that this graph was designed

to show change or growth, or pattern? _____

- - - - - - - - - - - - - - - - - - -

change or growth

12. The <u>supplementary amount-scale chart</u> is a chart that provides two kinds of measurement on a single chart.

First, it measures the variations in a series of data, just like any simple curve chart. In addition, it measures the size of this series in relation to another series.

For example, as shown in

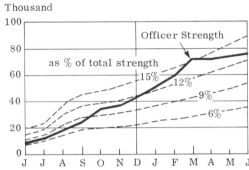

the chart above, a curve measuring officer strength (i.e., the number of officers) could be plotted against a series of supplementary curves showing 6%, 9%, 12%, and 15% of total military strength. Thus, although officer strength was plotted in <u>numbers</u> of officers in the Army at the end of each month of the year, we can get an idea of what <u>percentage of total Army strength</u> these numbers represent at any point in time (for that particular year, of course).

For instance, the chart tells us that in March there were about 70,000 officers in the Army, representing approximately 15% of total personnel strength.

About how many officers were there in December and what percentage of

total strength did this represent? _____

_____

- - - - - - - - - - - - - - - - - - -

somewhere in the vicinity of 42,000, representing about 11% of total Army personnel

13. The <u>index-scale chart</u> shows data that have been converted into percentages of a base value.

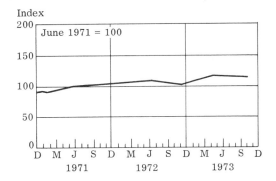

For example, production could be shown each month as a percentage of the level of June 1971. Thus, February (1972) might have a value of 102.3, March 105.7, etc. The principal use of index charts is to show <u>composite</u> data such as price indexes. However, they can be used also to compare two or more series of data that are measured in different kinds of units or in different size units.

While we're on the subject, let's note that an <u>index</u> is a purely arbitrary reference point selected to suit the comparison desired. For example, an established company with a slow but steady growth trend may have taken some vigorous action in 1965 to improve sales (initiated an extensive advertising campaign, introduced a new product, or something similar). They would like, of course, to know what effect this action has had on sales in subsequent years. Using 1965 sales of $3 million as a base line (index), they could then graph sales for future years simply as a percentage of increase (or decrease) over $3 million. Thus, if 1970 sales were $4.5 million, this would plot as a 50% increase. Or, if they let 1965 sales = 100 as a base (index), then sales for 1970 would plot as 150—a purely arbitrary number based on 1965 = 100.

A <u>composite</u> index is simply one that is made up of (that is, the combined average of) several individual components. A price index, for example, represents the average price of an array of consumer goods.

At the right is a popular index scale graph—the consumer price index (CPI). This index is used for the cost-of-living escalator in labor contracts (among other things) and reflects the rising cost of living that affects all of us. The graph uses the year 1967 as a base, that is, 1967 = 100. Although the calculation of this index is complicated, basically it represents a composite of the prices of a great many consumer items.

What was the approximate value of the index at the end of the year 1973?

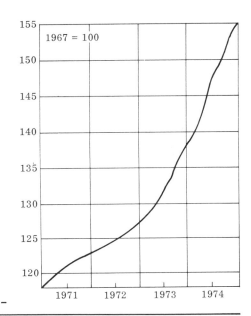

approximately 138

With this brief look at special scale graphs we conclude our overview of line graphs.

SELF-TEST

1. From the chart shown here would you say that the pattern of new incorporations is a smooth steady one on a month-to-month basis? Why? _____

_____

_____

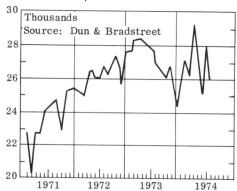

2. What conclusion could you draw about the relative movements of the Dow Jones industrial average (upper curve) and the NYSE composite index (lower curve)? _____

_____

_____

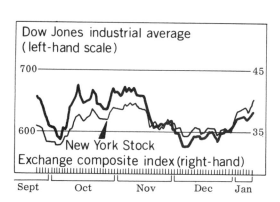

3.   What do you think is the purpose
of the chart at the right? That is,
what is it attempting to show?

_____

_____

_____

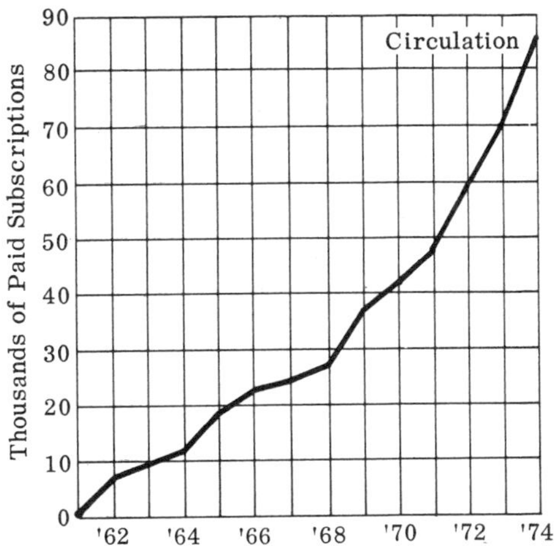

4.   Answer the following questions
about the graph at the right.

(a)   What is the general purpose of

the graph? _____

_____

(b)   What does the graph indicate regarding any apparently cyclical
tendency of the schedule through-

out the year? _____

_____

(c)   Would you say the deviations
from schedule pretty well aver-

aged out for the year? _____

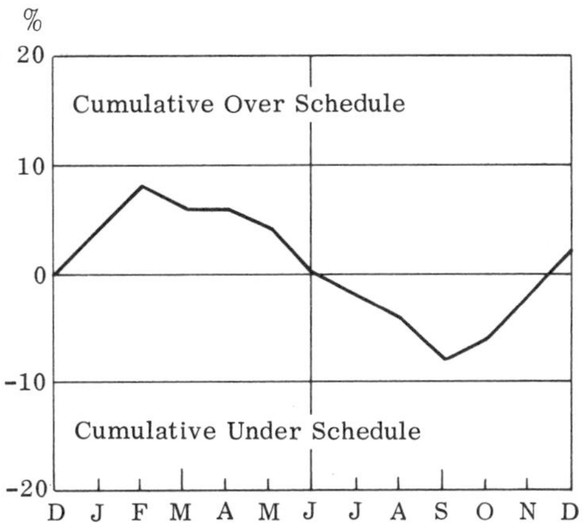

5.   Answer the following
questions.

(a)   What is the graph at
the right called?

_____

_____

(b)   What is its purpose?

_____

_____

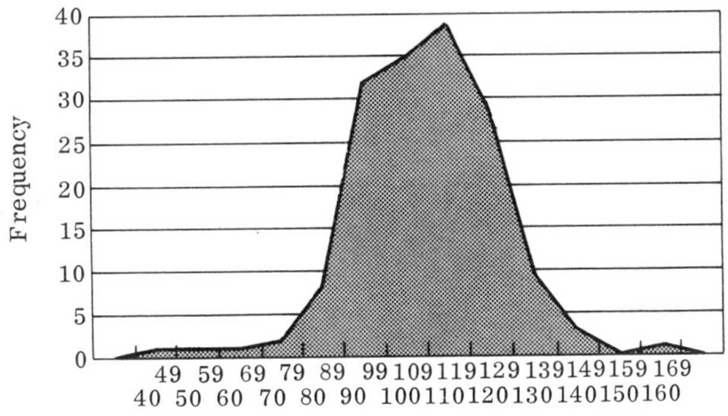

(c) What shape does the bottom curve on the previous page approximate?
That is, if the data base was larger (i.e., a larger sample), what would
the curve tend to become? _____

6. The graph at the right, known
as a _____,
shows the projected budget for a
company department. Do you think
a zigzag curve would be equally appropriate? Why? _____

_____

_____

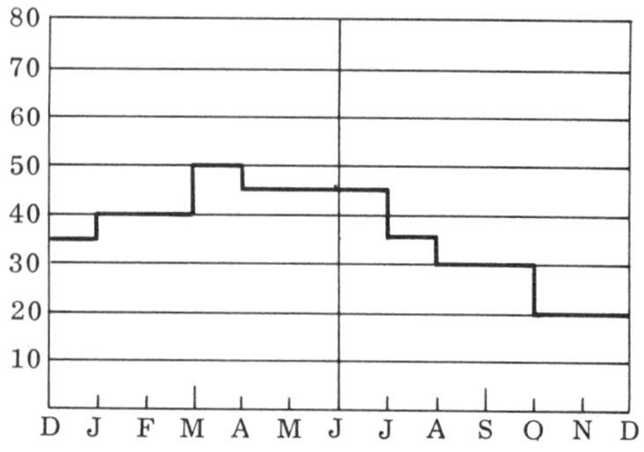

7. Assume you were the advertising manager of a large department
store and the sales manager handed
you the graph at the right representing his forecast of probable sales
levels of two departments during the
coming year.

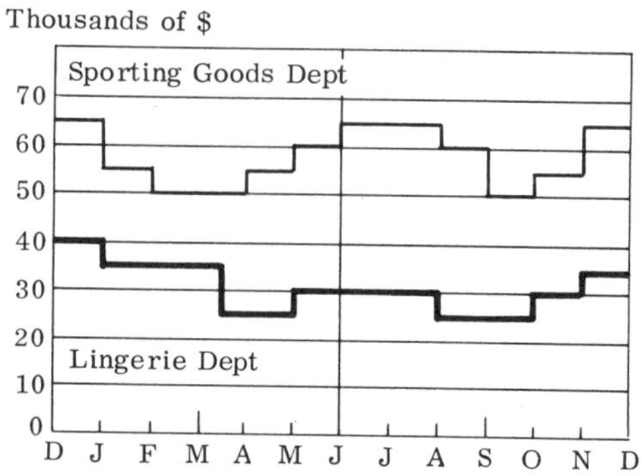

(a) What kind of sales pattern would
you say the sales manager was
predicting? _____

_____

(b) Which department do you think needs major advertising support?

_____

(c) What do you suppose accounts for the beginning, middle, and end highs
for the Sporting Goods department? _____

_____

8. If you were in the electrical ap-
pliances repair business and hired
help on an as-needed basis, what
would this chart showing number of
employees throughout the past 3
years tell you about:

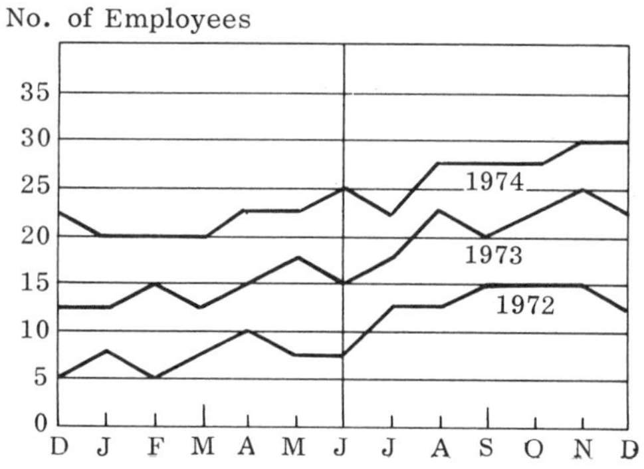

(a) The time of year when you need
most help? _____

_____

(b)   The trend of your help needs? _____

(c)   Could you safely assume from this chart (if you had no other indicators)

that your business is growing? _____

9.   The multiple time-scale chart
at the right shows the growth in
population of a fairly large western
city during two time periods, the
'50s and the '60s.

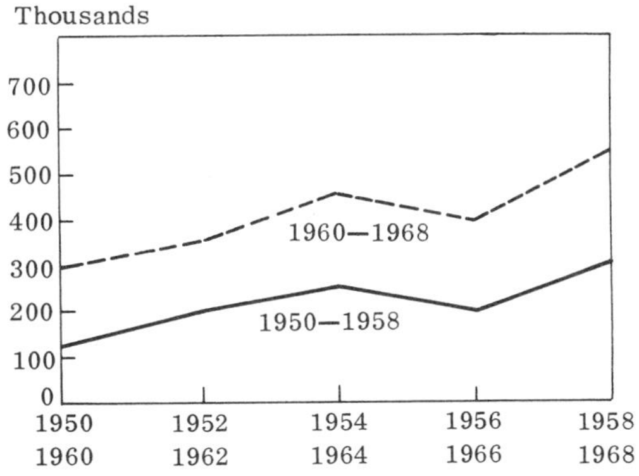

(a)   Is the pattern of growth fairly
consistent during both time

periods? _____

(b)   During which two-year period
was growth the strongest?

_____

(c)   Do you think the two matching periods of downward trend suggest anything

significant? _____

10. Answer the following
questions about the chart
shown here.

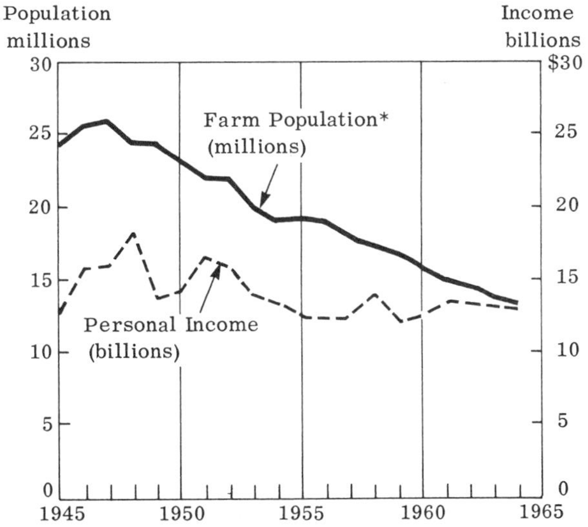

(a)   What two quantities
are being compared
in this chart?

_____

_____

(b)   What is happening to
farm population?

_____

_____

_____

(c)   What is happening to
personal income?

_____

_____

(d)   What is the significance of (b) and (c)? _____

_____

11. You will recognize the figure at the right as one that we
saw in frame 12. Let's dig into
it a bit further to make sure you
understand it. The total military
strength of the Army evidently
was changing (increasing) during
the period shown as indicated by
the fact that the percentage lines
representing numbers of enlisted
men and officers slope upwards.

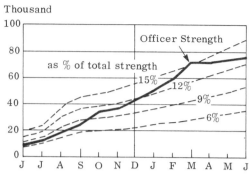

(a) If we assume a total mili-
tary strength of 500,000 in March, how <u>many</u> men does the 6% line repre-

sent at that time? _____

(b) How many men does 9% represent at that same time? _____

(c) Does the total number of men you found in answer to (b) agree with what
the chart shows? (Go straight up from March on the horizontal scale until you intersect the 6% line, then turn left and go straight across to the

vertical scale and see what it reads at that point.) _____

12. Shown at the right
is the ever popular con-
sumer price index—
this time for the Los
Angeles area. Notice
that it uses the prices
prevailing in 1967 as a
base.

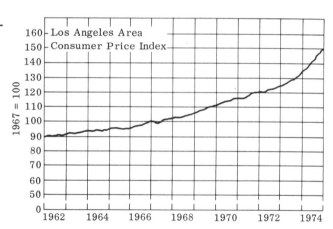

(a) What is the general trend of this index? _____

(b) What was the value of the index at the beginning of 1962?

_____

(c) What was the value of the index at the end of 1974? _____

<u>Answers to Self-Test</u>

1. No. Because the curve is very ragged, that is, it changes direction abruptly from month to month. (frame 1)

2. They moved generally together. (frame 2)

3. the cumulative increase of circulation of a newspaper or magazine, in thousands of paid subscriptions, since 1961 (frame 3)

4. (a) to show cumulative deviations from schedule in percent
    (b) It appears generally to run ahead of forecast in the first half of the year and behind forecast the last half. Thus, it may be cyclical, completing a single cycle throughout a period of one year. (We'd have to see the graph for several years to see if it was consistently cyclical.)
    (c) Yes, that is the clear visual impression the chart gives.
    (frame 4)

5. (a) a frequency polygon
    (b) to display the frequency distribution of data samples
    (c) It tends to approximate a normal curve; the rough similarity is apparent already.
    (frame 5)

6. Simple step curve. No, because zigzag curves usually reflect abrupt, month-to-month changes and budget forecasts generally are not of this kind, extending instead over a period of time. (frame 7)

7. (a) Larger scales of sporting goods than lingerie; strongest sales for both departments at Christmas plus another strong rise for sporting goods during summer; a bottoming out of sales in spring and fall for both departments.
    (b) Lingerie, both because its sales are consistently lower but also because it needs some kind of strong support during the middle of the year.
    (c) Probably Christmas and summer buying, as suggested in (a) above.
    (frame 8)

8. (a) You need <u>most</u> during the second half of the year.
    (b) They are steadily increasing.
    (c) You <u>might</u> infer that but, without other supporting indicators, you could not assume it—certainly not from this chart alone. Your supply of competent workers might be diminishing, and inefficient help might mean that you are having to hire more people to do the same (or perhaps slightly more) work. Inefficiency certainly could disguise your true rate of growth.
    (frame 9)

9. (a) Yes, it appears to be, since the two curves show a nearly perfect matching pattern.

(b) 1960-1968 (a growth of 250,000 compared with about 175,000 for the period 1950-1958).

(c) They <u>suggest</u> that there may be some common cause for this matching decline in the growth pattern, and that it might be worthwhile to discover what that cause is. But by themselves the curves do not provide any clues as to the reason.
(frame 10)

10. (a) Farm population and personal income.
   (b) It is declining.
   (c) It is, on the average, remaining almost steady.
   (d) One interpretation might be that, with farm population decreasing and (total) personal income staying fairly steady, individual income is increasing since a fewer number of people are generating the same volume of income dollars.
   (frame 11)

11. (a) $.06(500,000) = 30,000$
    (b) $.09(500,000) = 45,000$
    (c) yes
    (frame 12)

12. (a) A steady increase both in the <u>amount</u> and in the <u>rate</u> of increase of consumer prices. (Notice that during the first $5\frac{1}{2}$ years the index rose 10 points, in the next $4\frac{1}{2}$ years it rose 20 points, and in the final 3 years reported it rose 30 points, hence the accelerating rate.)
    (b) 90
    (c) nearly 150

CHAPTER FOUR
Area Graphs and Bar Graphs

Thus far in our investigation of graphs we have been concerned only with the shape of the curve represented. Now we are going to consider some graphs in which we are interested, either primarily or secondarily, in the significance of the area under the curve.

When you have reached the end of this chapter, you will be familiar with the construction, significance, and purpose of area graphs and bar graphs. Specifically, you will learn how to recognize, interpret, and use:

- area graphs (including calculation of areas);

- the family of vertical bar graphs, including simple, grouped, subdivided, deviation, floating, and range-column charts;

- the family of horizontal bar graphs, including simple bar, bar-and-symbol, subdivided bar, 100% bar, grouped bar, paired bar, deviation bar, sliding bar, range-bar, change-bar, and progressive bar charts.

AREA GRAPHS

1. We learned in Chapter 1 (frame 20) that surface graphs are simply shaded line graphs. Some serve the same general purpose as similar line graphs, and some can be used in special ways. In addition to the four types of surface graph we looked at in Chapter 1—simple zigzag, simple step, subdivided zigzag, and curvilinear—there are these additional varieties: subdivided step, 100% surface, and the band graph. But because the surface graphs are so closely associated with the line graphs, and because shading serves mainly to assist the reader visually in distinguishing between the plotted lines, we will not devote additional space to them here.

However, what is important is that surface charts not be confused with area charts. They are not the same thing, even though they look alike. An example should help make the distinction.

In both surface graphs and line graphs each plotted point represents a linear distance (i.e., measurement in one direction). Thus as shown at the right each point on the zigzag curve represents some discrete value of the vertical scale at a given point in time, and <u>there is no functional relationship between the vertical and horizontal quantities</u>. Shading the space below the curve

produces a shaded area—but not an area chart. For it to be defined as an area chart, the two plotting variables would have to be <u>functionally</u> related in some way. That is, it should be possible to express their relationship by an equation of some sort.

Let's take a case in which the variables <u>are</u> functionally related. In the graph at the right notice that the two variables are time and speed. Therefore, if we plot a point representing a speed of 50 mph for a period of 3 hours, we in effect establish the rectangular area shown whose size is 3 × 50 = 150. But since we know that speed mul-

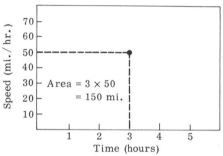

tiplied by time is distance, we realize that the figure of 150 is the graphic equivalent of 150 <u>miles</u>.

This area happened to be rectangular and, therefore, easy to calculate. But regardless of the area shape, its value can be determined by the product of some vertical value and some horizontal value. This therefore relates the area to the name and calibration of the vertical and horizontal axes. This is true even though the area is so irregular that the only way it can be determined is by counting the number of graph-paper squares bounded by the line or curve.

Suppose, for example, we wished to know the area under the curve between the two abscissa (horizontal) values 4 (seconds) and 9 (seconds). How would you go about this? We have suggested one way, can you think of others?

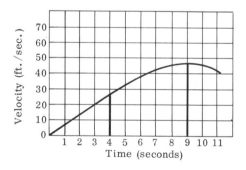

(1) You could count the squares under the curve, guessing at the areas of those that were split by the curve.

(2) You could use a device known as the <u>planimeter</u>, which is a device for accurately measuring plane areas of any form by simply moving a tracing pin aroung the periphery of the figure and reading from the measuring dial.

(3) You could use the method of area evaluation known as the Trapezoidal Rule, which in effect reduces every area to a collection of regular figures, regardless of the shape of the curve. Basically this involves:

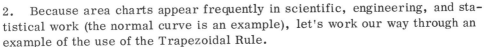

(a) Subdividing the area into several trapezoids (as shown at the right).

(b) Determining the area of each of the trapezoids.

(c) Adding together the areas of the trapezoids. This sum is the value of the area in question.

(4) If you knew the <u>equation</u> of the curve you could, by the methods of integral calculus, calculate the area under the curve between the selected limits.

2. Because area charts appear frequently in scientific, engineering, and statistical work (the normal curve is an example), let's work our way through an example of the use of the Trapezoidal Rule.

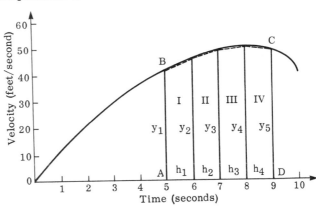

The curve at the right represents the velocity-versus-time plot of an object under the influence of several forces. Velocity is the vertical dimension and time the horizontal dimension.

Since we know that the product of velocity and time represents distance, our object will be to determine graphically the distance covered by this object from the 5-second time point to the 9-second time point. The area we are concerned with, then, is the area labeled ABCD in the figure above. If this were a regular area such as a rectangle or a triangle we would have no difficulty in evaluating it. But since the upper boundary of the area is curvilinear, no simple formula for area will work. Hence we will apply the trapezoidal rule.

The first step is—as we learned in the answer to frame 1—to subdivide the area involved into several trapezoids. (Why trapezoids? Because the

trapezoid is a regular geometrical figure whose area can be easily computed.)
We do this by drawing vertical lines between the horizontal base line and the
curve at the 5-, 6-, 7-, 8-, and 9-second time points, thus dividing the area
into four areas, identified by the Roman numerals I, II, III, and IV. The question of how many trapezoids to use is a trade-off between how accurate you
wish to be and how much work you're willing to do in the way of calculation.
The more trapezoids, the more accuracy—but the more arithmetic, also.

A trapezoid has two parallel sides, and in this case these consist of the
vertical lines marked y_1, y_2, y_3, y_4, and y_5. A third (nonparallel) side in
each case is the section of the base line identified as h_1, h_2, h_3, or h_4. The
fourth (also nonparallel) side is drawn so that it closely approximates the
curve (shown as a dashed line).

Thus by using several trapezoids we have managed to subdivide the area
under the curve into several regular geometrical figures that closely follow
the curve. You can see from the diagram that only a very small error is involved if we assume that the sum of the areas of the four trapezoids is equal
to the area bounded by the curve arc BC, the ordinate lines y_1 and y_5, and the
portion of the horizontal base line between 5 and 9 seconds. Notice also that
each of the four trapezoids has equal bases—that is, the lengths of lines h_1,
h_2, h_3, and h_4 are all equal. This is done merely to simplify the calculation
of the total area, a calculation we are now ready to perform.

The area of a trapezoid equals half the sum of the bases multiplied by the
altitude. Thus, the area of each trapezoid can be expressed in equation form
as follows.

$$\text{Area I} = \tfrac{1}{2}(y_1 + y_2)h_1$$
$$\text{Area II} = \tfrac{1}{2}(y_2 + y_3)h_2$$
$$\text{Area III} = \tfrac{1}{2}(y_3 + y_4)h_3$$
$$\text{Area IV} = \tfrac{1}{2}(y_4 + y_5)h_4$$

Since all of the h values are the same, the sum of these four areas, A_t,
can be expressed as:

$$A_t = h(\frac{y_1}{2} + \frac{y_2}{2} + \frac{y_2}{2} + \frac{y_3}{2} + \frac{y_3}{2} + \frac{y_4}{2} + \frac{y_4}{2} + \frac{y_5}{2})$$

But since $\frac{y_2}{2}$, $\frac{y_3}{2}$, and $\frac{y_4}{2}$ each appear twice, we can restate the expression for
total area as:

$$A_t = h(\frac{y_1}{2} + y_2 + y_3 + y_4 + \frac{y_5}{2})$$

In other words, to find the total area it is necessary merely to add the sum of
the middle bases, plus half the first and last bases, and multiply the total by
any one of the equal altitude values.

Referring to our figure we can read the (approximate) values of the bases
as:

$y_1 = 41$ ft/sec, hence $\dfrac{y_1}{2} = 20.5$ ft/sec

$y_2 = 45$ ft/sec
$y_3 = 48$ ft/sec
$y_4 = 51$ ft/sec

$y_5 = 49$ ft/sec, hence $\dfrac{y_5}{2} = 24.5$ ft/sec

$h = 1$ sec

Therefore $A_t = 1(20.5 + 45 + 48 + 51 + 24.5) = 189$ What does the 189 represent? How do we know this? _____

- - - - - - - - - - - - - - - - - - - -

It represents distance in feet that the object travels during the 5th to 9th seconds. We know this by the following arithmetic reasoning: Since the area represents distance, then

$$A_t = \text{Distance} = 1 \; \text{sec} \left(189 \; \frac{\text{ft}}{\text{sec}}\right)$$

$$\text{Distance} = 189 \text{ ft}$$

VERTICAL BAR GRAPHS

3.   In Chapter 1 we saw that there are two main kinds of bar charts: vertical (or columnar) and horizontal bar. Remember that many people use the term bar to mean only a horizontal bar, and the term columnar (or just column) to mean a vertical bar chart. Therefore we will use the words column and vertical bar interchangeably.

In bar charts, amounts are represented by the height or length of the bars. Column charts generally are used to compare data for a given item at different times. Horizontal bar charts (which we will discuss a little later) generally are used to compare data for different items at the same time.

Column charts are more suitable for data on activities that occur during a period of time (such as sales during June) than for figures indicating status on a given date (such as inventory on June 30). Figures indicating status may be shown as column charts but usually they can be presented better by linear or surface graphs.

A column chart makes a stronger picture than a line graph showing the same data and is especially good for showing growth or development in a striking manner. Also, columns usually are better than line graphs when there are only three or four time points to be plotted or when the series fluctuates very sharply. They are not as good for comparing several series of data, for showing a long series of data with a great many plottings, or for showing numerous components of totals.

Having considered some general characteristics of vertical bar graphs let's look now at some specific types.

The <u>simple column chart</u> consists of a series of vertical bars, each extending from the base line to a particular vertical height. Like all such charts it must have a zero line.

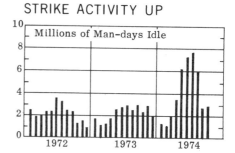

STRIKE ACTIVITY UP

The example at the right shows one way in which columnar charts can be used to good effect. This particular chart shows the monthly level of strike activity, in millions of man-days idle, over a period of nearly three years. Here we have data on one item (time lost) compared at different times.

Each column in this chart represents a month of the year—January through December—although they are not designated. In which month (and year) was

strike activity (and time lost) the greatest? _____

- - - - - - - - - - - - - - - - - - - -

July 1974

4. A logical variation of the simple column chart is the <u>grouped column chart</u>, which consists of groups of two or more vertical columns and is useful for comparing two (but seldom more) series of data. It is also known as a compound column chart.

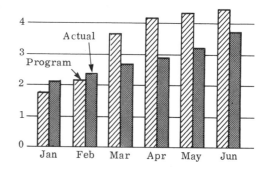

This type is most effective when used for series that differ in level or in trend. To set off each group of columns from the next, the spacing between groups should be at least as wide as a column. The two sets of columns should have shadings that contrast sharply in tone.

Sometimes this type of chart is drawn with one column in each pair overlapping the other. But this treatment should be used only when needed to save space and when the front set of columns always is shorter than the back set.

Referring to the graph above, how would you analyze what is happening to actual costs as compared with program (estimated) costs? (The significance of the vertical scale numbers is not important for the purpose of this analy-

sis.) _____

_____

- - - - - - - - - - - - - - - - - - - -

Actual costs ran slightly higher than program costs during January and February but since then have been running consistently less—suggesting improved program control.

5. The <u>subdivided column chart</u> is used to show the component parts of a series of totals. It is also known as a segmented column chart.

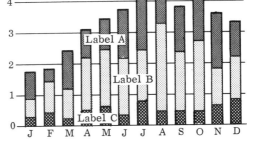

This chart serves essentially the same purpose as the subdivided surface chart but is a better choice when the plotted values fluctuate sharply from one period to the next.

Notice that by pushing the columns together and leaving only the vertical connections between column ends, we obtain a step or multiple-step graph.

Inexperienced chartmakers often try to show too many components in this type of chart. More than three or four segments are difficult to identify and to compare. If a large number of component parts must be shown, the subdivided surface chart usually is a better choice.

As a general rule it is not a good idea to connect segments in adjoining columns by diagonal lines as such lines usually do more to confuse than to clarify.

At the right is a subdivided column chart showing the component groups, by age, comprising the total unemployed for each of the years indicated.

(a) Approximately how many 16- to 19-year-olds were unemployed during 1958? _____

(b) Approximately how many 25- to 54-year-old females were unemployed during the year 1971? (This will involve a little subtraction.)

_____

- - - - - - - - - - - - - - - - - - - -

(a) about 600,000; (b) about 3,300,000 minus 2,300,000 or approximately 1 million

6.   Another variety of vertical
bar chart that often comes in
handy is the <u>deviation-column
chart</u>. It is used to show the differences between two series, and
usually presents negative values
plotted below the zero line as well
as positive values above.

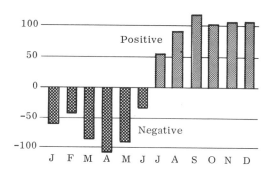

In a deviation chart a column
can extend either above or below
the base line, but not both. This
type of graph is good for showing
how much results deviated from an estimate or requirement. It is especially
useful for measuring net gains and losses and all similar plus-or-minus differences. In working with this or any other type of deviation chart you should
be aware that positive deviations are usually used for "good" results and negative deviations for "bad" results.

Below is an example of a deviation-column chart that can't fail to impress
any of us, considering the subject matter.

In 16 Years: 15 Budget Deficits

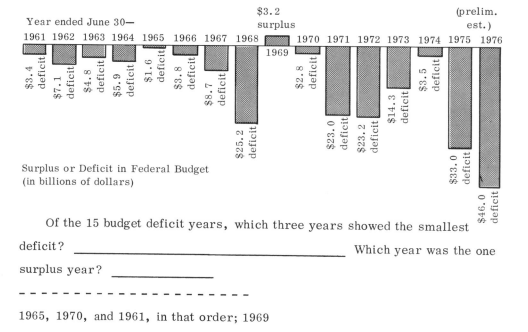

Of the 15 budget deficit years, which three years showed the smallest
deficit? _____ Which year was the one
surplus year? _____

- - - - - - - - - - - - - - - - - - - -

1965, 1970, and 1961, in that order; 1969

7.   The <u>floating column chart</u> is a special variation of the subdivided column
type. The total length of the column represents the total of two main components, one of which is plotted above the zero line and the other below.

In the example shown at the right, the total length of the column might represent the total number of contracts completed during a month, the portions above the base line indicating the number completed on or ahead of schedule and the portions below the base indicating those completed behind schedule.

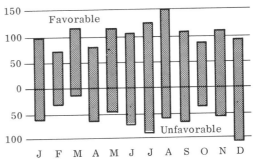

This type of chart differs from the deviation type in that each column may extend both above and below the zero line. In effect, floating columns are subdivided columns with the dividing line between segments used as a base. The overall length of the column thus is subordinated to the comparison of the segments.

The chart at the right probably will look familiar since it represents part of an ad that appears frequently in newspapers and magazines. Which of the four oils compared are they trying to tell us

is the best one? _____

_____

– –

Safflower oil, of course, because it is "highest in polyunsaturates." Whether you knew or cared about polyunsaturates you would know that safflower oil was the intended winner because the maximum portion of its column is on the upper (positive) side of the base line.

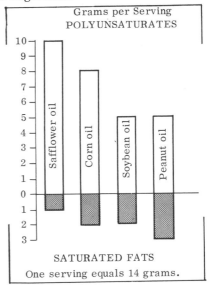

8. The last type of vertical bar chart we will consider is the so-called <u>range-column chart</u>. This chart emphasizes the difference between two sets of values. The highest and lowest value during each period are connected by a column measuring the high-low range.

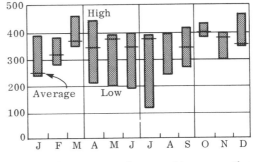

In this type of columnar chart supplementary information, such as the average value, can readily be included by means of cross-lines or other indicators. This chart is useful for showing the spread in such data as personnel strength, inventories, and price ranges.

As with all charts measuring range this type can be used effectively to show a supplementary range such as high and low tolerance limits, upper and lower levels of efficient operation, or other top and bottom "bench" marks. These are usually put in as light dash or dot lines across the entire chart.

INDUSTRIALS

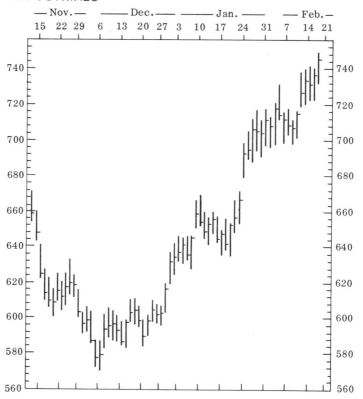

If you read the Wall Street Journal the chart above should be familiar to you since it appears daily. It is one of a series showing the trading range (range of buying and selling prices) of several groups of selected stocks—industrials, transportation, and utilities. The top of each bar represents the highest price paid for the industrials stock group on any particular day, the bottom of the bar the lowest price, and the "tick" mark in between shows the closing price.

Using a straightedge to help, see if you can determine the high, low, and closing prices for the group on January 10. _____

- -

High—approximately 668; Low—approximately 652; Closing—approximately 654

(Note: For simplicity we have referred to these figures as "prices." Actually they represent index values <u>related</u> to the selling/buying prices but including some other factors as well. Our only purpose at this point is to learn how to read this type of chart. However, if you are interested in learning more about how the various indices are arrived at, the New York Stock Exchange has free pamphlets available that explain such matters. Or of course any broker will be glad to explain it to you.)

HORIZONTAL BAR GRAPHS

9.   Horizontal bar graphs may very well be the type of graphs with which you are most familiar since they appear so frequently in reports, magazines, statistics for popular consumption, and financial summaries.

In a bar chart, such as the one at the right, amounts are represented by the length of horizontal bars. Throughout this section, where we refer to bar charts you will know we mean <u>horizontal</u> bar charts. Bar charts differ from line, surface, and column charts in that <u>they have only one scale</u>.

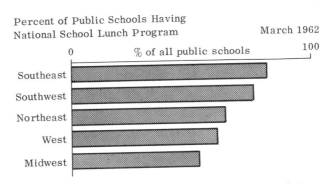

This is an <u>amount</u> scale (percent in the figure above) and it measures across the chart. The vertical dimension is used for listing the items measured (public school areas of the United States in the above example). The horizontal direction of the bars presents a direct contrast to the vertical arrangement of columns and thus tells you at a glance you are not looking at a time series chart.

All bar charts (except the range-bar chart which we will meet later) need a zero line or other base line. The bars should not be broken except to avoid a long "freak" bar, that is, one that is entirely out of scale with the others.

Bar charts can be used effectively to show how several items differ from one another in one or two characteristics, or to show how several items differ from one another in the distribution of their components.

The <u>order</u> of the items in a bar chart is important. Many kinds of items can be listed in several different ways, each giving a different emphasis to the data. The <u>thickness</u> of the bars has no measurement value and so should be uniform throughout the chart.

The <u>simple bar chart</u> is merely a series of horizontal bars drawn to the right of a common base line. Each item can be plotted according to its absolute value or expressed as a percentage of an appropriate total (as we saw in the example at the beginning of this frame).

The graph at the right is a combined bar chart and table of values from which it was derived. Referring to the income gap graph, in which year(s) did the median income of blacks reach the highest percentage of that of whites? What has the percentage been doing since then?

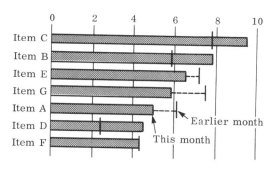

The Income Gap

Percent of Black to White

Year	Percent of Black to White	Median Income (Blacks)	Median Income (Whites)
1964	50%	$3,724	$6,858
1965	50%	3,886	7,251
1966	58%	4,507	7,792
1967	59%	4,875	8,234
1968	60%	5,360	8,937
1969	61%	5,999	9,794
1970	61%	6,279	10,235
1971	60%	6,440	10,672
1972	59%	6,864	11,549

- -

In the years 1969 and 1970 it reached 61%; since then it has been receding.

10. The <u>bar-and-symbol chart</u> presents a simple bar comparison with a secondary comparison added.

The additional data, such as budgeted amounts, forecasts, averages, plans, or other standards greatly increase the value of the chart. The symbols also can be used to show data for some earlier time. For example, the preceding month or year, or the best previous performance. This type of chart can, therefore, often be used to summarize several time series charts.

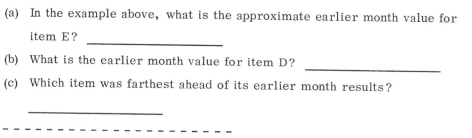

Different symbols can be used to emphasize different comparisons. Thus, to emphasize the bar comparison, use vertical cross-lines (as shown above) or some other subordinate device such as a hollow diamond.

(a) In the example above, what is the approximate earlier month value for item E? _____

(b) What is the earlier month value for item D? _____

(c) Which item was farthest ahead of its earlier month results?

_____

- -

(a) about 7.2; (b) about 2.4; (c) It looks like a close tie between items C, B, and D, but D is the winner.

11. The <u>subdivided bar chart</u> shows each bar divided into its component parts. It is used to show clearly how much effect each component has on the size of the total.

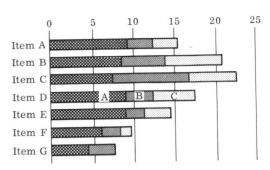

Usually the largest or most important component is put next to the zero line. As in subdivided column or surface charts, only the component that starts from the base can be measured directly from the scale. Thus, in the example above, to find the value of component A for item C we must read (estimate) the value on the horizontal scale opposite the end of the A component; this appears to be about 7. But to find the value of component C we need to read its end value (or total bar value) on the horizontal scale (about 22.5) and subtract from it the value of the end of component B (about 17). This tells us that the value of the C component is approximately 22.5 minus 17, or 5.5.

(a) Following this same general procedure find the value of the B component for item C. _____

(b) Which component (A, B, or C) would you say has the greatest variation throughout the seven items? _____

- - - - - - - - - - - - - - - - - - - -

(a) B value minus A value equals B value. Thus, 17 – 7 = 10, the value of the B component.

(b) B

12. The <u>100% bar chart</u> shows each bar divided according to the relative size of its components, regardless of the absolute size of the total value of the bar. This chart emphasizes the proportionate part of the total contributed by each component.

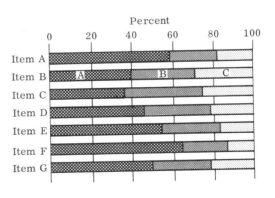

This form of subdivided bar graph has the advantage of having two base lines, zero and 100%, and so provides direct comparison of the components at either end of the chart. Like other percentage comparisons it should be used with caution when there is a wide difference in the <u>absolute</u> amounts on which it is based.

In working with this chart keep in mind that the components (identified in the above example as A, B, and C) must add up to 100%. Thus if we wish to know what portion (percentage) of the total of item G each of the three components represents, we first read the value of component A, which turns out to be approximately 50%. Next, reading the endpoint value of component B (76%) and subtracting from this the value of component A (50%) gives us a value of 26% for B. The value of component C can be found simply by reading backward from the 100% line to the endpoint of component B, or subtracting the endpoint reading of component B from 100%. Either method gives us a value of 24% for component C.

Use this procedure to find the values of the three component parts of item A.

A = _____

B = _____

C = _____

- -

A = about 58%;   B = about 24%;   C = about 18%

13. The <u>grouped bar chart</u> permits the comparison of a number of items in two respects at the same time.

Although the bar-and-symbol treatment can be used to present this kind of information, grouped bars are better (when space permits) because they suggest the nature of the comparison better. Groups of more than two or three

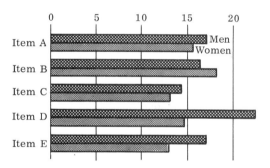

bars seldom are desirable, however, because the eye is confronted with too many comparisons at once.

In designing this chart the category of major interest usually is put first and is given the darker shading. For spacing, the standard is to make the space between groups no less than the thickness of a bar. This allows the eye to separate the groups rather easily while still perceiving the implied association of the groups.

The sample graph above shows a comparison between men and women relative to five different items. These items could be such things as longevity, earning power, digital dexterity, or IQ, or they might represent a series of anthropometric characteristics, such as height, weight, arm reach, torso length, and so forth. The point is that we could have as many comparison items as we wished. This is different from, for instance, a grouped column chart in which we seldom have more than three comparison factors.

Referring to the graph on the previous page:

(a) On which item did women receive their highest score? _____

(b) On which item did women rank higher than men? _____

(c) On which item did men receive their lowest score? _____

- - - - - - - - - - - - - - - - - - - -

(a) B;   (b) B;   (c) C

14. The <u>paired bar chart</u> provides
another way to compare a number
of items in two respects. It is pre-
ferable when <u>different units or
scales must be used for each cate-
gory</u>.
     Notice that in this type of chart
instead of being grouped the bars
are placed opposite one another,
one set extending to the left of the
item names and the other set to

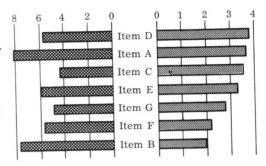

the right. In the example above, the bars at the left might represent cost per
unit and those at the right number of units produced. Or the number of motor
vehicle accidents per 100,000 miles and the number of miles operated.

     Bars extending to the left (like columns extending below the zero line)
tend to suggest unfavorable results or conditions. If the nature of the data
permits, therefore, the chart should be designed to take advantage of this.

     Charts of this kind commonly are arranged so that on one side the bars
appear in order of decreasing magnitude (length). On which side of the above

chart has this been done? _____

- - - - - - - - - - - - - - - - - - - -

On the right side. This is done to give emphasis to the comparison factor considered most important. (That's why, in this case, the items are not in alpha-
betical order.)

15. The <u>deviation-bar chart</u> presents bars extending left or right,
each item having only one bar. It
is useful for comparing differences between actual results and a
program standard, especially
when these differences are so
small that they would be hard to
compare on a chart showing the
totals.

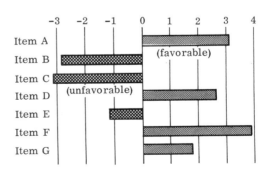

The sample chart at the beginning of this frame might represent any kind of "more-than-less-than" data such as amount over or under allowance, percent over or under estimate, or amount over or under budget. Like any deviation chart this type is effective also for showing other net measurements such as net gains and losses or net ins and net outs.

In how many instances do the favorable (positive) items exceed a value of

+2? _____ Or the unfavorable (negative) items a value of -2?

_____ On the whole would you say that the positive factors out-

weigh the negative ones? _____

- -

three; two; yes—not just because there are <u>more</u> of them, but because their combined <u>length</u> certainly would exceed that of the combined negative bars

16. The <u>sliding bar chart</u> is a special kind of subdivided bar chart in which the length of the bar represents the total of two main components, one to the left and one to the right of the base line.

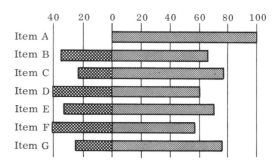

For example, personnel might be divided into military on one side of the base line and civilian on the other. Or the quantity of an item in storage divided into ready-for-issue and not-ready.

The main advantage of this sliding treatment is that both of the components can be measured directly from a common base. This type of chart is, therefore, primarily a comparison of <u>parts</u>, with the comparison of totals secondary. A conventional subdivided bar chart, on the other hand, is primarily a comparison of <u>totals</u>.

Referring to the example above:

(a) In how many cases was the component on the left greater than the component on the right? _____

(b) Which item was comprised of only one component? _____

(c) Are the bars on the right arranged in order of decreasing magnitude?

_____

- -

(a) none; (b) item A; (c) no, because this might give undue emphasis to the right hand component

17. The <u>range-bar chart</u> shows the range or spread between low and high amounts, rather than just the size of single amounts.

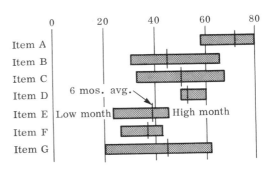

Range bars do not start from a common base and so cannot be compared directly. Instead they show the size of each range with relation to the two amounts it separates (i.e., the highest and lowest figures). This type of presentation is useful when there is interest in the amounts at each end of the range as well as in the difference between them. By adding a cross-bar (or some similar device) to a simple range chart, it can be used to show a comparison of averages, plus the range of values behind these averages.

For example, the chart above might represent the range of monthly sales (in dollars) of various models of cars by an automobile dealer. Thus, throughout the period of six months the lowest monthly sales for item A (a particular car model) amounted to about $57,000 and the highest $80,000, the <u>average</u> monthly sales for the six months being about $72,000.

Assuming that the figures in the chart represent automobile sales in units of a thousand dollars:

(a) Which item (car model) had the highest average sales? _____

(b) Which had the lowest month's sales? _____

(c) Which had the least variation in monthly sales volume? _____

(d) Which had the greatest variation in monthly sales volume? _____

- - - - - - - - - - - - - - - - - - - -

(a) A; (b) G; (c) D; (d) G

18. The <u>change-bar chart</u> is similar to the range-bar chart except that arrowheads have been added to the bars to show <u>change</u> instead of simple range.

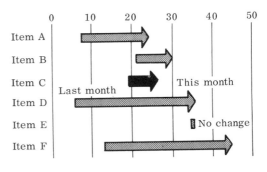

The addition of the pointer not only tells which end of the bar marks the current result, but also tells whether the change is an increase or a decrease. <u>Unfavorable</u> changes (whether decreases <u>or</u> increases) usually are shown as black bars to make them stand out clearly from favorable changes. This type of presentation is not necessarily limited to showing change from one time to another. The bars might show such information as original and revised programs for various items, or performance before and after adoption of new methods or procedures.

Notice in the chart at the beginning of this frame that item D shows a last-month reading of 5 at the left end of the bar, and a this-month reading of about 35 at the right (arrowhead) end, whereas item E is unchanged at a value of 35.

What values do you get for item F? This month _____. Last

month _____. What kind of change does item C represent?

- -

this month = 45; last month = 12; an unfavorable increase

19. The <u>progressive bar chart</u>
(also known as a step-by-step bar
chart) is an unusual form of subdivided bar chart. It shows the
segments of a bar as a series of
steps, one below the other, thus
calling attention to <u>one step at a
time</u> and emphasizing the number of steps involved.

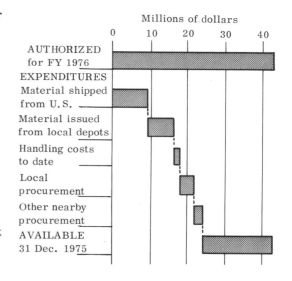

Putting the pieces one under
another in this way also permits
easier and fuller labeling which
is especially helpful when labels
are long, descriptive captions.
The step-by-step method can be
used either to build up or to break
down a total. For example, it
could show how several sources
of funds add up to the total available, or how the total available has been used.

The dashed vertical lines between segments simply serve to convey to the reader the fact that the ending point of one segment is the beginning point of the next. In other words it is simply a graphic convention to assist visual communication.

The total (AUTHORIZED) bar at the top is broken into six segments. Using a straightedge to help you read the horizontal scale values, see if you can determine the approximate values (in millions of $) of each of the six segments.

1st _____   2nd _____   3rd _____

4th _____   5th _____   6th _____

- -

1st = 9; 2nd = 8; 3rd = 1; 4th = 4; 5th = 2; 6th = 19

Now it's time to review what we have covered on the subject of area charts, and vertical and horizontal bar charts. The following quiz should <u>help you find out</u> what topics you may need to review.

SELF-TEST

1.   In the area chart the variables are related in what way? _____

2.   An area chart and a surface graph are the same thing but with different names. (True or False) _____

3.   The curve at the right is the familiar power curve we first saw in frame
15 of Chapter 2. Apply the Trapezoidal
Rule to approximate the area under the
curve between x = 1 and x = 4 and indicate what it represents. (Notice that
this is a velocity-versus-time plot like
the one we worked with in frame 2.)
Here, again, is the formula:

$$A_t = h(\frac{y_1}{2} + y_2 + y_3 + \frac{y_4}{2})$$

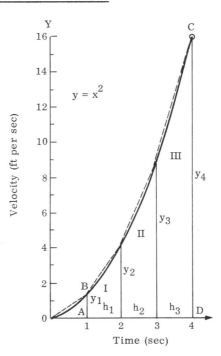

4.   Looking at this simple column chart
showing the world population growth,
would you conclude that the population is
growing at a steady (i.e., linear) rate?

Explain. _____
_____

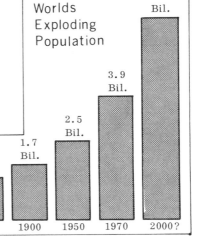

5. The graph at the right is a type of grouped column chart. One obvious difference from the example we saw in frame 4 is the fact that the pairs of columns don't actually touch.

(a) What is the vertical scale?

_____

(b) Were U.S. exports to Latin America greater in 1961 than in 1957, or less?

_____

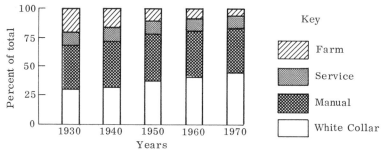

Latin American Trade with U.S.

Latin American ☐
Rest of the world ▨   Key

Total U.S. exports   Total U.S. imports

24.2% 17.3%   29.0% 22.7%

75.8% 82.7%   71.0% 77.3%

1957   1961   1957   1961

(c) In 1957 were U.S. imports from the rest of the world (i.e., other than Latin America) greater than its exports to the rest of the world?

_____

6. Answer the questions which follow the graph below.

Percent Distribution of Occupations

Key

▨ Farm
▨ Service
▨ Manual
☐ White Collar

(a) What type of graph is shown? _____

(b) Is the percentage of people engaged in farming increasing or decreasing?

_____

(c) What would you conclude about the percentage of those engaged in manual occupations between 1930 and 1970? _____

_____

_____

7.   Answer the following questions about the graph at the right.

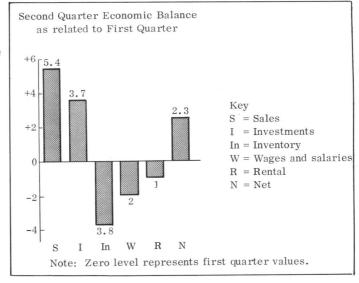

Second Quarter Economic Balance as related to First Quarter

Key
S = Sales
I = Investments
In = Inventory
W = Wages and salaries
R = Rental
N = Net

Note: Zero level represents first quarter values.

(a) How would you identify the graph? _____

(b) What does the zero line represent? _____

(c) What is being compared? _____

(d) Were wages and salaries higher or lower in the second quarter than in the

first? _____

8.   The floating column chart at the right shows the totals of new and used car sales on a monthly basis for one full year of operation of an automobile agency. What conclusion can you draw with regard to:

(a) New car sales? _____

_____

_____

(b) Used car sales? _____

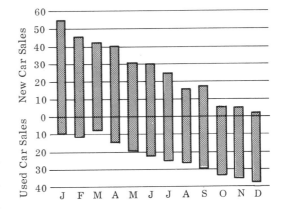

_____

(c) Total sales? _____

9. What does the range-column chart at the right tell you generally about the pattern of prices received by farmers for feed grains during the period reported?

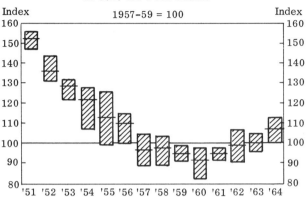

High-Low Prices Received by Farmers in U.S. for Feed Grains

1957-59 = 100

10. Judging by the chart at the right:

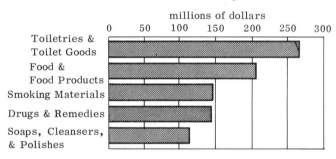

Leading Television Advertisers
United States, 1965

millions of dollars

(a) How do TV advertising expenditures for toiletries and toilet goods compare with those for soaps, cleansers, and polishes? _____

(b) Does the pattern of advertising expenditures suggest anything to you about the margin of profit in a product line? _____

11. This chart presents a breakdown of Army civilian personnel overseas, by commands. Would you say that personnel levels in the various commands generally have decreased or increased since March 31?

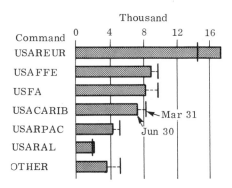

12. Referring to the subdivided chart at the right:

Materials Purchased for Insulation in Standard and Compact 1961 Passenger Cars

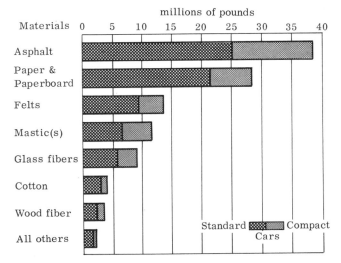

(a) Which material was used in approximately equal amounts in the standard and compact cars? _____

(b) How many million pounds of glass fibers were used for standard cars?

_____

(c) How many pounds of cotton were used in compact cars? _____

13. The graph at the top of the next page shows a distribution of vehicles disabled (for the reasons indicated) during a particular month.

(a) Which service had the highest
     percentage of disablements due
     to lack of parts?

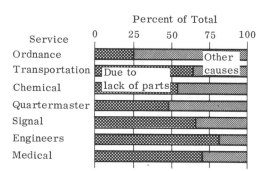

_____

(b) Which had the highest disable-
     ments due to other causes?

_____

14. If in this chart A represents
personnel discharges and B represents enlistments:

(a) Is personnel strength gener-
     ally increasing or decreas-
     ing?

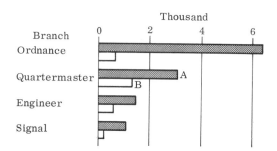

_____

(b) Which branch had the most
     enlistments?

_____

15. The paired bar chart below reflects the alumni annual support program
for a major university in a particular year.

ANNUAL SUPPORT PROGRAM

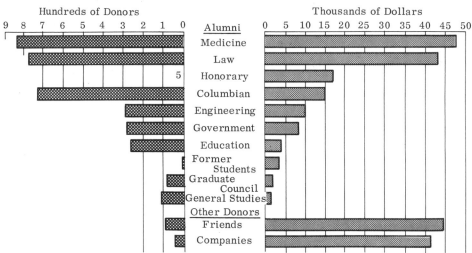

(a) Was the amount of financial support from any one group of donors gener-
     ally proportional to the number of donors? _____

(b) Are there any significant exceptions to (a)? Which? _____

16. From the chart at the right what general trend do you observe in the change in industrial life insurance in force during the ten-year period covered?

_____

_____

_____

_____

_____

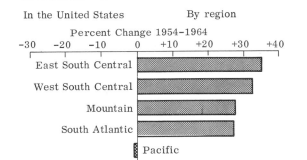

INDUSTRIAL LIFE INSURANCE IN FORCE

In the United States   By region

Percent Change 1954-1964

17. This range-bar chart shows the percentage of vehicles disabled four days or more due to lack of parts over a period of six months.

(a) What is the average figure for ordnance vehicles?

_____

(b) Which service had the smallest spread between high and low months?

_____

(c) Which service had the greatest spread? _____

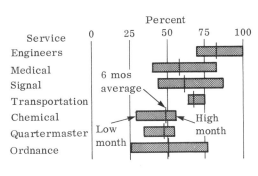

18. The progressive bar chart at the right presents the income of U.S. life insurance companies for a particular year. Based on what it shows:

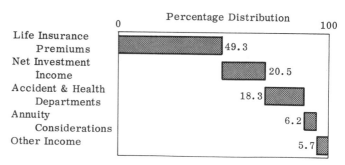

(a) About one-half of the income comes from what single source?

_____

(b) About what <u>fraction</u> of the total income does investment income represent?

_____

Answers to Self-Test

1. functionally; that is, the relationship between the two variables can be expressed by an equation (frame 1)

2. False (frame 1)

3. $y_1 = 1$, $y_2 = 4$, $y_3 = 9$, $y_4 = 16$

   $A_t = 1(\frac{1}{2} + 4 + 9 + \frac{16}{2})$

   $A_t = 21\frac{1}{2}$ (the actual value, found by the methods of calculus is 21)
   (frame 2)

4. No, it is not growing at a steady (constant) rate. If it were, a line connecting the top-center points of the seven columns would be straight. But if you can visualize such a line you will see that it would, in fact, be curved—curving rapidly upward, showing an <u>accelerating</u> rate of growth. (frame 3)

5. (a) percent; (b) less; (c) no, less
   (frame 4)

6. (a) subdivided column (vertical bar) chart
   (b) decreasing
   (c) It is remaining fairly constant.
   (frame 5)

7. (a) as a deviation-column chart
   (b) first quarter values
   (c) second quarter results with first quarter results
   (d) lower
   (frame 6)

8. (a) They decreased during the year.
    (b) They increased during the year.
    (c) They diminished somewhat.
    (All of these may be seasonal.)
    (frame 7)

9. They decreased rather steadily from 1951 to 1960 and then moved upward
    gradually to 1964, the last year shown. (frame 8)

10. (a) They are more than twice as large.
    (b) It suggests that manufacturers can afford to spend more more money
    for advertising products such as toiletries because they represent a
    greater margin of profit. But of course this is not necessarily true since
    many other factors are involved.
    (frame 9)

11. Generally speaking (in five out of seven instances), they have decreased.
    (frame 10)

12. (a) mastic(s)—or possibly wood fiber; (b) about 6 million; (c) about one
    million
    (frame 11)

13. (a) Engineers; (b) Ordnance
    (frame 12)

14. (a) decreasing; (b) Quartermaster
    (frame 13)

15. (a) Yes, generally speaking.
    (b) Indeed yes! While among the very smallest in numbers of donors,
    Friends and Companies were among the very highest in thousands of dol-
    lars contributed. And the five honorary alumni also contributed a dispro-
    portionate amount.
    (frame 14)

16. Generally, industrial life insurance in force tended to increase in the
    southern and mountain regions and to decrease in the northern and New
    England regions. (frame 15)

17. (a) 50%; (b) Transportation; (c) Ordnance
    (frame 17)

18. (a) life insurance premiums
    (b) one-fifth (20.5%)
    (frame 19)

CHAPTER FIVE
Special Graphs

Now we come to the general lumping together of what we have simply termed special graphs. They are special only in the sense that they don't fit into any of the categories of graphs we have covered thus far—line graphs, surface and area graphs, and bar graphs. Specifically, we are going to discuss why, when, and how to use these special graphs to meet special needs, and what they should convey to us as readers when we come across them.

When you have arrived at the end of this chapter you will be able to:

- recognize, interpret, and use pictographs (or pictograms), histograms, circle graphs (or pie charts), scattergrams, and nomographs (or nomograms);

- do simple curve fitting for empirically derived data.

1. The pictograph (or pictogram) is a graphic representation of a statistic using small, simple pictures to represent a certain number or amount. It is essentially a variation of the bar chart in which a row of representative symbols is used to make up the bar.

In the example shown here, the purpose of the chart is to portray statistics relating to the increasing number of individuals 65 and older who form part of the population of the United States. Obviously some of the figure-symbols, each of which represents 2 million persons, have to be split down the middle in some fashion to represent less than 2 million. However, in most pictographs numbers are

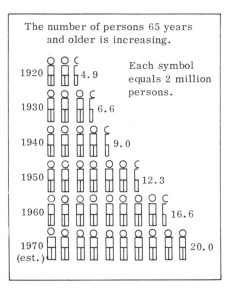

The number of persons 65 years and older is increasing.

Each symbol equals 2 million persons.

1920 4.9
1930 6.6
1940 9.0
1950 12.3
1960 16.6
1970 (est.) 20.0

added at the end of each row to specify the precise quantity the symbols are supposed to represent.

Although pictographs are seen frequently in reports, they do have drawbacks. One which we have just observed is the difficulty of representing numerical amounts accurately by the use of symbols. Also, the pictograph costs more to produce, is much harder to do well and, most important, is very limited in the kinds of comparison it can present.

Pictograph charts almost always turn out to be bar charts and, as we have seen, many kinds of statistical comparison cannot be shown satisfactorily in that form. Pictorial treatments are best used for "popularizing"—drawing the attention of readers who have little or no interest in the subject or knowledge of the data.

Consider the following pictograph in the light of our discussion above.

 First Grade Pupils Enrolled
in Public Elementary Schools

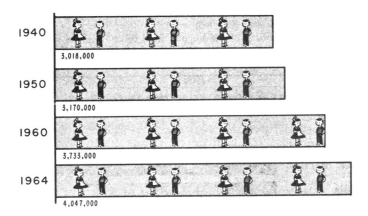

(a) What similarities or differences do you find in comparison with the earlier pictograph shown in this frame? _____

_____

(b) What quantity of pupils is each pair of figures intended to represent?

_____

- -

(a) Although the figures here are used to represent quantity (of first grade pupils) symbolically, the chartist has relied on the length of the bar in which they are contained to show the quantity more precisely. In the earlier pictograph no actual bar was drawn and the chartmaker has relied upon the figures themselves (sometimes sliced to suit the data) to show quantity visually. In both cases the quantities also are given in numbers.

(b) 1 million

2.   A <u>histogram</u> is a columnar graph in block form made by plotting frequency
of occurrence against the values obtained.   It is the common mode of representing frequency of distribution in statistics.

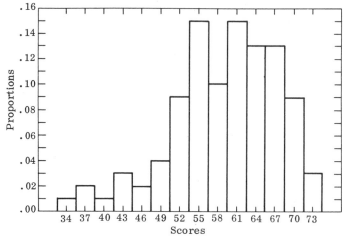

Notice that removing the vertical lines that extend all the way to the base
line—leaving the connections between levels—gives a step graph.   However,
there is a very important reason for those vertical lines in a histogram.   Each
column of the histogram represents a <u>range</u> of scores, and the columns must
adjoin one another because the range of scores is continuous.   Thus, the first
column represents scores from 33 to 35, the next column scores of 36 to 38,
the next 39 to 41 and so on.   This is quite different from a columnar graph,
for example, where each column represents a different—and not necessarily
related—quantity.

You may recall that in frame 5 of Chapter 3 we discussed a chart known
as a frequency polygon and that its definition was very similar to that given
above for the histogram.   The reason for this is that, although of different
form, they portray essentially the same information.   To obtain the curve
shown in frame 5 (Chapter 3) we used the data from columns 1 and 2 of the table shown in frame 10 of Chapter 2.   Mark your place here and refer back to
the table and frequency polygon on pages 69 and 91, to refresh your memory
before going on.

Notice that the frequency polygon represents a plot of scores against the
frequency with which they occurred, that is, the data from columns 1 and 2 of
the table.   We could, just as easily, have plotted this data in the form of a
histogram.   And if, instead of using <u>frequency</u> for our ordinate values we instead plot <u>proportion</u> (from column 3) against scores, we obtain the histogram
shown in this frame.

The proportion values are, as you will recall, obtained simply by dividing
the frequency values by 158, the total number of scores in the sample.   Since
the frequency and proportion values are essentially equivalent, our histogram
would look about the same regardless of which we used.   Actually, the usual
sequence in statistical work is to plot the sampling data in the form of a frequency distribution histogram first and then derive the frequency polygon from
it, if required.

What principal difference do you observe between the appearance of the histogram (as shown in the example in this frame) and the columnar charts we have studied? _____

_____

- -

There is no space between the columns in the histogram; that is, they adjoin one another. In columnar charts there is always some space (usually the width of a single column) between the columns.

3. The circle graph (or sectogram) is a pictorial representation of a complete circle, or "pie," that is sliced into a number of wedges, the size of each showing its percentage of the entire quantity.

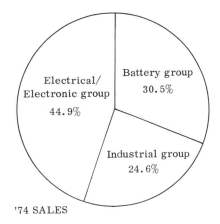

The example at the right is adapted from an advertisement aimed at showing the changing emphasis in the company's sales to electrical/electronic items. Thus, although a similar graph for the prior year was not shown, we are meant to assume that the "slice of the pie" was smaller for electrical and electronics products previously (and proportionately larger for the other company product lines). This is a typical use of circle graphs.

The circle graph, or pie chart, appears to be simple and "nonstatistical," so it is a popular form of presentation for general readers. However, since the eye can compare linear distances more easily and accurately than angles or areas, the component parts of a total usually can be shown more effectively in a chart using linear measurement.

Pie charts are awkward to label and do not fit as well on a report page as bar comparisons (vertical or horizontal). Thus a series of pies is less effective than a series of subdivided bars (or columns) for comparing a group of subdivided totals. Several pies require much more space than several bars. Moreover, the comparable components often are in a different location in each pie and so are hard to compare.

Nevertheless, circle graphs have their uses, especially to show sources of income and modes of expenditure for governmental agencies, corporations, and nonprofit charitable organizations that depend upon public solicitation for their funding.

The chart shown at the top of the next page is intended to show citizens where their State tax monies go. See if you can turn this into a bar chart. You will have five items (bars) and your scale will be a percentage scale, which need be no longer than the largest percentage figure shown in the graph.

Where the Money Goes

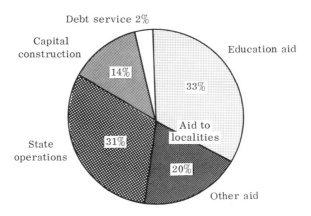

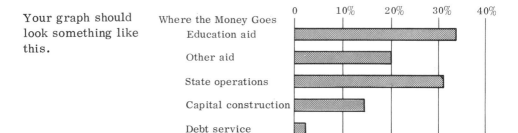

Your graph should
look something like
this.

4.   Now we're going to move on to a statistical type of graph known as a
scattergram. A scattergram, or scatter diagram, is a plot of points representing a series of observed relationships between two variables.

This type of graph is used frequently by researchers, hence the variables might be any of a great many
things—height and weight, family income versus number of children,
scores on two different tests, or
others. Suppose, for example, we
were interested in knowing if rich families tend to have many children or
few children, or if there is no relationship at all between wealth and the number of children in a family. We make
a number of observations (counts) of
the number of children in families of
varying degrees of wealth and then
plot them in a graph like the one shown.

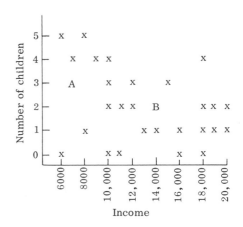

Each x on the scattergram represents a family in our sample. The height of the x on the graph represents the number of children in the family. The left-right position of the x represents family income. For example, A represents a family with three children and an income of $7000. What does B represent? _____

- - - - - - - - - - - - - - - - - - - -

a family with two children and an income of $14,000

5. The objective in using a scattergram is to try to perceive some trend. Looking at the graph shown at the right, would you say that (in this sample) families with high incomes tend to have relatively larger or smaller families?

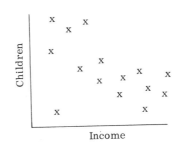

_____

- -

smaller

6. What we would like to be able to find is some kind of a function—straight or curvilinear—that would provide a reasonably good relationship between two measurements. Study the diagram at the right carefully and then see if you can draw one straight line that you feel fits reasonably well the central trend of the data points (x's).

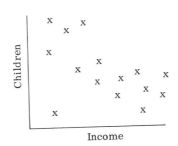

- -

Hopefully your straight line looks something like the one shown at the right. Obviously this is a matter of judgment, and the more experience you have in doing it the better your judgment will become. Nevertheless, it is apparent that there is some kind of linear relationship between the two variables.

7. You have just been engaged, in a small way, in what is called curve fitting. When there is no relationship between the two variables your x's will be scattered randomly over the graph area. They will not, therefore, show any clear linear, or curvilinear, relationship.

The stronger the relationship between the measures, the closer together the x's will fall, forming either a linear or a curvilinear pattern.

In the space provided under each of the following drawings, write in the name you feel best describes the pattern of data points found there—linear, curvilinear, or random.

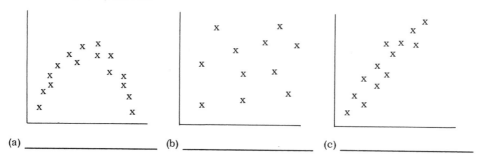

(a) _____   (b) _____   (c) _____

- - - - - - - - - - - - - - - - - - - -

(a) curvilinear; (b) random; (c) linear

8.   Scattergrams are simply plots of empirical data, that is, data derived from <u>observation</u>. And the purpose in using a scattergram is to try to perceive some <u>trend</u> between the two variables involved. If we can represent such a trend by, say, a straight line, then we have found something that may enable us to predict relationships between the variables for other values that we did not observe.

In frame 10 the overall appearance of the data points <u>suggested</u> there might be a linear relationship and you tried your hand at drawing a straight line that seemed to fit that relationship. However, because they are based on individual interpretation, lines drawn by inspection are subject to wide error, hence statisticians usually calculate them—something we will not attempt to do here.

Shown at the right is a typical scatter plot of data from two variables (tests, actually).

(a)   Observing the data points, do you feel there is a relationship between the two variables? Why?

_____

_____

_____

(b)   Could a relationship best be represented by a curve or a straight

line? _____   _____

(c)   Draw a trend line.

- -

(a) There seems to be such a relationship since the directional trend of the points is from the lower left corner of the chart towards the upper right.
(b) By a straight line.
(c) (See figure at right.)

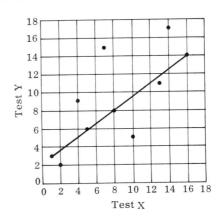

9. The last type of graph we are going to consider is the <u>nomograph</u>. The nomograph, or <u>nomogram</u>, is a graph that enables the user with the aid of a straightedge, to read the value of a dependent variable when the values of two or more independent variables are given.

The example shown next is a wind chill nomogram. The left vertical scale represents wind velocity in both miles per hour and meters per second. The scale at the right is a temperature scale, calibrated both in degrees Fahrenheit and centigrade. The sloping scale is the wind chill scale. To use this nomogram you first find the temperature and wind velocity values that interest you. Then with a straightedge laid across these two values, you can read the wind chill condition where the straightedge crosses the wind chill scale.

Nomographs have been developed to assist engineering and scientific work in an almost limitless number of areas. A great many of these have been published in <u>Design News</u> magazine (Cahners Publishing Company) during the past ten or more years, and many others are available in various collections of engineering and design data.

Using a straightedge to assist you, what wind chill reading do you get for:

(a) A temperature of 0° centigrade and a wind velocity of 1 mile per hour?

_____

(b) A temperature reading of 0° centigrade and a wind velocity reading of

10 mph? _____

(c) Has the increase in wind velocity made much of a change in the wind chill

reading? _____

- - - - - - - - - - - - - - - - - - - -

(a) Essentially, "Cold."
(b) Ranges from "Very cold" through "Bitterly cold" and the following three extreme categories.
(c) Yes; it has made a severe change from simply uncomfortable cold to a condition of real danger from short exposure.

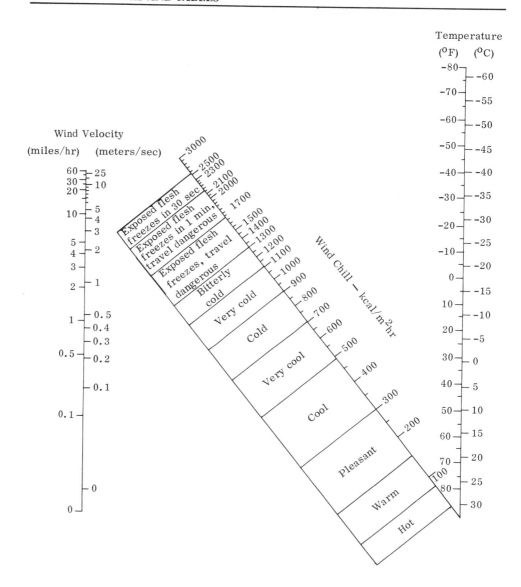

10. Although not really a nomograph, the <u>conversion scale</u> is at least a near relative. Instead of being composed of two independent variables (scales) and one dependent variable (scale) like the nomograph, the conversion scale (or <u>chart</u>, since it also can be presented in chart form) has just two scales. <u>Either</u> scale can be considered the independent—or dependent—variable depending upon what information you are seeking.

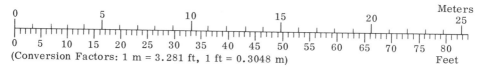

(Conversion Factors: 1 m = 3.281 ft, 1 ft = 0.3048 m)

For example, the scale across serves to convert meters to feet or feet to meters. Thus if you wanted to know how many feet there are in 15 meters, you read along the upper (meters) scale to the number 15, and below it, on the feet scale find the value $49\frac{1}{2}$. This tells you that 15 meters is equivalent to about $49\frac{1}{2}$ feet.

Conversely, to convert 600 feet to meters, find the number 60 on the lower (feet) scale, read 18.2 (meters) on the upper scale, and multiply by 10 to give a value of approximately 182 meters.

Use this conversion scale to find:

(a) How many feet there are in 25 meters. ____________

(b) How many meters there are in 40 feet. ____________

- -

(a) about 83 feet; (b) about 12 meters

SELF-TEST

1. Which of the two pictographs at the right do you think is best for displaying accurate information about bread sales? Why?

____

____

____

____

____

____

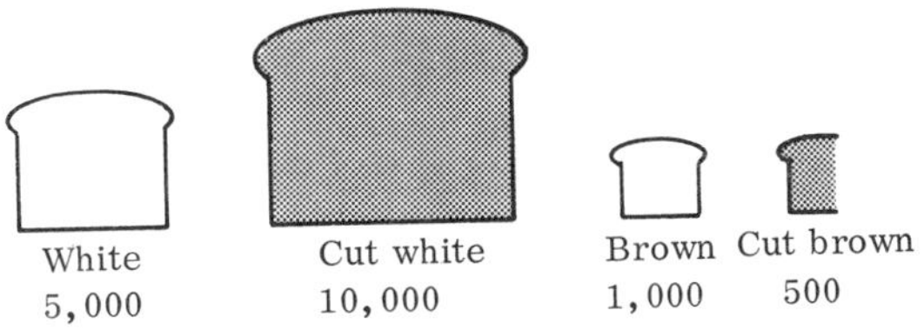
White
5,000
Cut white
10,000
Brown
1,000
Cut brown
500

Pictograph A

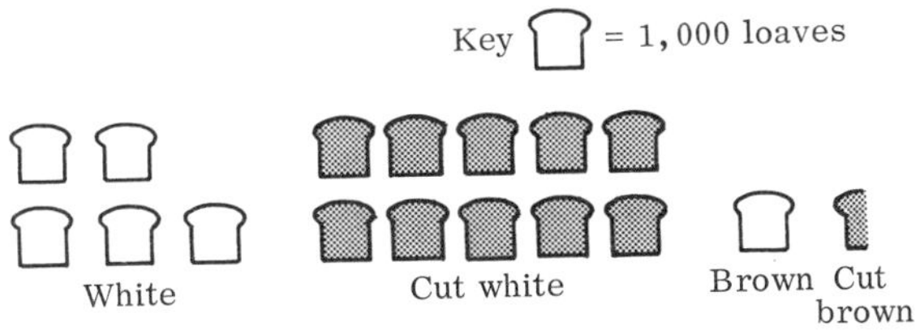
Key
= 1,000 loaves
White
Cut white
Brown
Cut brown

Pictograph B

2. Is the graph at the right a valid example of a histogram? Explain your answer.

____

____

____

____

____

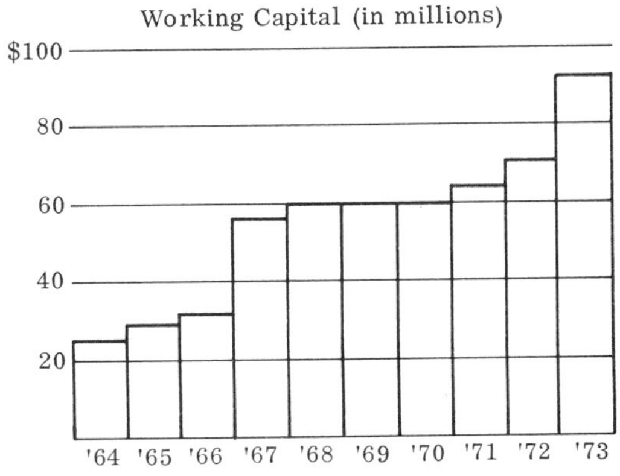
Working Capital (in millions)
$100
80
60
40
20
'64
'65
'66
'67
'68
'69
'70
'71
'72
'73

3.   Is the graph shown at the right a valid example
of a circle graph? Explain your answer.

_____

_____

_____

SOHO
SQUARE

39.8   27.7

1962 1973

4.   (a)   What is a graph like the one at the right

called? _____

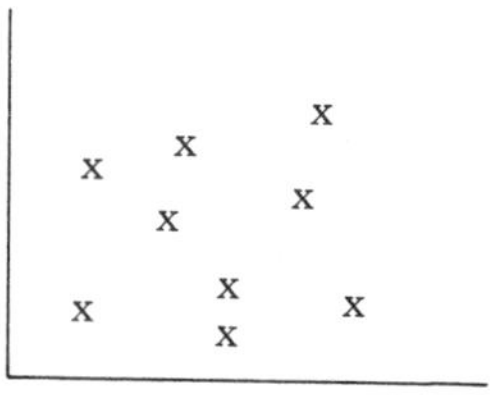

(b)   From what you see here would you or would
you not suspect that there is a significant relationship between the two variables (meas-

ures)? Why? _____

5.   Which of the graphs
at the right shows the
strongest relationship
between two variables?
Why?

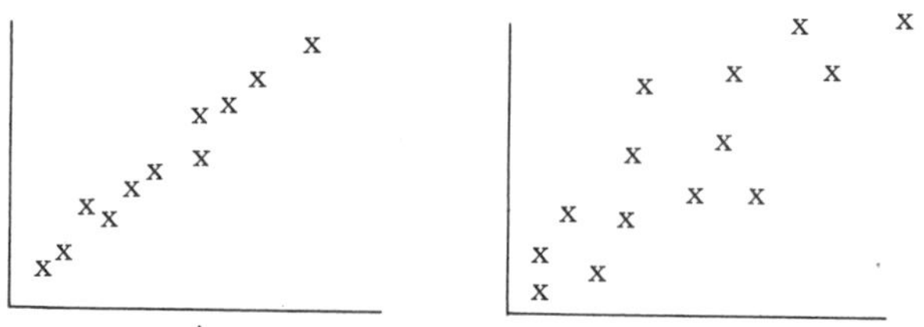

_____

_____

A

B

6. Mach number is a measure of speed. It is the ratio of the speed of an airplane (in mph) to the speed of sound (also in mph) at any particular altitude. Use the nomograph at the right to find the Mach number of an airplane that is flying at an altitude of 20,000 feet and a speed of 1000 mph.

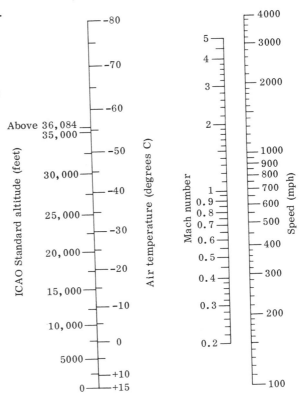

Answers to Self-Test

1. Pictograph B. Because in working with pictographs it is important that all symbols be the same <u>size</u>, otherwise we get the type of misleading area distortion we discussed in Chapter 1. (frame 1)

2. No. It is not a histogram because it does <u>not</u> represent frequency distribution. True, the columns adjoin one another as in a histogram, but this is not the only feature of a histogram nor the most important one. Actually, this would be known as a connected column chart. (frame 2)

3. No. It is not a circle graph at all because the circle has not been subdivided into sectors to represent parts of the whole. The circle has simply been used to draw attention to the two columns representing the values being compared. (frame 3)

4. (a) a scattergram, or scatter diagram
   (b) No, because there is no apparent trend or pattern in the data points. (frames 4, 5, and 6)

5. A because the data points show a stronger trend than those of B. (frames 7-8)

6. about 1.3 (frame 9)

A PARTING WORD

With the completion of this Self-Test you have arrived at the end of the chapter and of the book.

As stated at the outset the purpose throughout has been to help you learn how to interpret graphs and tables. Hopefully, we have succeeded. To assist you in doing so we have, as it were, gone behind the scenes and given you a glimpse into the methods of construction and basic purposes of a number of specific types of graphic representation. This approach stems from the conviction that only if you really understand the basic concept underlying a chart or table can you interpret intelligently the information it seeks to convey.

Finally, as stated earlier, although we have tried to include representative examples of the most common types, there is a nearly limitless variety of formats for plotted and tabularized data. What you have learned here should, however, enable you to analyze, categorize, and interpret most of the graphs and tables you encounter in your work or reading.

By applying what you have learned here, your appreciation of the ability of graphs and tables to present complex information in a clear and concise way will continue to grow. And so, of course, will your ability to extract that information.

Appendix

The pages which follow contain the table of common logarithms, which you were referred to in the book. You will also find a table which classifies the various charts and graphs covered in this book, according to form or type and according to the chapter in which each category is discussed.

COMMON LOGARITHMS OF NUMBERS*

N	0	1	2	3	4	5	6	7	8	9
10	0000	0043	0086	0128	0170	0212	0253	0294	0334	0374
11	0414	0453	0492	0531	0569	0607	0645	0682	0719	0755
12	0792	0828	0864	0899	0934	0969	1004	1038	1072	1106
13	1139	1173	1206	1239	1271	1303	1335	1367	1399	1430
14	1461	1492	1523	1553	1584	1614	1644	1673	1703	1732
15	1761	1790	1818	1847	1875	1903	1931	1959	1987	2014
16	2041	2068	2095	2122	2148	2175	2201	2227	2253	2279
17	2304	2330	2355	2380	2405	2430	2455	2480	2504	2529
18	2553	2577	2601	2625	2648	2672	2695	2718	2742	2765
19	2788	2810	2833	2856	2878	2900	2923	2945	2967	2989
20	3010	3032	3054	3075	3096	3118	3139	3160	3181	3201
21	3222	3243	3263	3284	3304	3324	3345	3365	3385	3404
22	3424	3444	3464	3483	3502	3522	3541	3560	3579	3598
23	3617	3636	3655	3674	3692	3711	3729	3747	3766	3784
24	3802	3820	3838	3856	3874	3892	3909	3927	3945	3962
25	3979	3997	4014	4031	4048	4065	4082	4099	4116	4133
26	4150	4166	4183	4200	4216	4232	4249	4265	4281	4298
27	4314	4330	4346	4362	4378	4393	4409	4425	4440	4456
28	4472	4487	4502	4518	4533	4548	4564	4579	4594	4609
29	4624	4639	4654	4669	4683	4698	4713	4728	4742	4757
30	4771	4786	4800	4814	4829	4843	4857	4871	4886	4900
31	4914	4928	4942	4955	4969	4983	4997	5011	5024	5038
32	5051	5065	5079	5092	5105	5119	5132	5145	5159	5172
33	5185	5198	5211	5224	5237	5250	5263	5276	5289	5302
34	5315	5328	5340	5353	5366	5378	5391	5403	5416	5428
35	5441	5453	5465	5478	5490	5502	5514	5527	5539	5551
36	5563	5575	5587	5599	5611	5623	5635	5647	5658	5670
37	5682	5694	5705	5717	5729	5740	5752	5763	5775	5786
38	5798	5809	5821	5832	5843	5855	5866	5877	5888	5899
39	5911	5922	5933	5944	5955	5966	5977	5988	5999	6010
40	6021	6031	6042	6053	6064	6075	6085	6096	6107	6117
41	6128	6138	6149	6160	6170	6180	6191	6201	6212	6222
42	6232	6243	6253	6263	6274	6284	6294	6304	6314	6325
43	6335	6345	6355	6365	6375	6385	6395	6405	6415	6425
44	6435	6444	6454	6464	6474	6484	6493	6503	6513	6522
45	6532	6542	6551	6561	6571	6580	6590	6599	6609	6618
46	6628	6637	6646	6656	6665	6675	6684	6693	6702	6712
47	6721	6730	6739	6749	6758	6767	6776	6785	6794	6803
48	6812	6821	6830	6839	6848	6857	6866	6875	6884	6893
49	6902	6911	6920	6928	6937	6946	6955	6964	6972	6981
50	6990	6998	7007	7016	7024	7033	7042	7050	7059	7067
51	7076	7084	7093	7101	7110	7118	7126	7135	7143	7152
52	7160	7168	7177	7185	7193	7202	7210	7218	7226	7235
53	7243	7251	7259	7267	7275	7284	7292	7300	7308	7316
54	7324	7332	7340	7348	7356	7364	7372	7380	7388	7396
N	0	1	2	3	4	5	6	7	8	9

* This table gives the mantissas of numbers with the decimal point omitted in each case. Characteristics are determined by inspection from the numbers.

COMMON LOGARITHMS OF NUMBERS (cont.)

N	0	1	2	3	4	5	6	7	8	9
55	7404	7412	7419	7427	7435	7443	7451	7459	7466	7474
56	7482	7490	7497	7505	7513	7520	7528	7536	7543	7551
57	7559	7566	7574	7582	7589	7597	7604	7612	7619	7627
58	7634	7642	7649	7657	7664	7672	7679	7686	7694	7701
59	7709	7716	7723	7731	7738	7745	7752	7760	7767	7774
60	7782	7789	7796	7803	7810	7818	7825	7832	7839	7846
61	7853	7860	7868	7875	7882	7889	7896	7903	7910	7917
62	7924	7931	7938	7945	7952	7959	7966	7973	7980	7987
63	7993	8000	8007	8014	8021	8028	8035	8041	8048	8055
64	8062	8069	8075	8082	8089	8096	8102	8109	8116	8122
65	8129	8136	8142	8149	8156	8162	8169	8176	8182	8189
66	8195	8202	8209	8215	8222	8228	8235	8241	8248	8254
67	8261	8267	8274	8280	8287	8293	8299	8306	8312	8319
68	8325	8331	8338	8344	8351	8357	8363	8370	8376	8382
69	8388	8395	8401	8407	8414	8420	8426	8432	8439	8445
70	8451	8457	8463	8470	8476	8482	8488	8494	8500	8506
71	8513	8519	8525	8531	8537	8543	8549	8555	8561	8567
72	8573	8579	8585	8591	8597	8603	8609	8615	8621	8627
73	8633	8639	8645	8651	8657	8663	8669	8675	8681	8686
74	8692	8698	8704	8710	8716	8722	8727	8733	8739	8745
75	8751	8756	8762	8768	8774	8779	8785	8791	8797	8802
76	8808	8814	8820	8825	8831	8837	8842	8848	8854	8859
77	8865	8871	8876	8882	8887	8893	8899	8904	8910	8915
78	8921	8927	8932	8938	8943	8949	8954	8960	8965	8971
79	8976	8982	8987	8993	8998	9004	9009	9015	9020	9025
80	9031	9036	9042	9047	9053	9058	9063	9069	9074	9079
81	9085	9090	9096	9101	9106	9112	9117	9122	9128	9133
82	9138	9143	9149	9154	9159	9165	9170	9175	9180	9186
83	9191	9196	9201	9206	9212	9217	9222	9227	9232	9238
84	9243	9248	9253	9258	9263	9269	9274	9279	9284	9289
85	9294	9299	9304	9309	9315	9320	9325	9330	9335	9340
86	9345	9350	9355	9360	9365	9370	9375	9380	9385	9390
87	9395	9400	9405	9410	9415	9420	9425	9430	9435	9440
88	9445	9450	9455	9460	9465	9469	9474	9479	9484	9489
89	9494	9499	9504	9509	9513	9518	9523	9528	9533	9538
90	9542	9547	9552	9557	9562	9566	9571	9576	9581	9586
91	9590	9595	9600	9605	9609	9614	9619	9624	9628	9633
92	9638	9643	9647	9652	9657	9661	9666	9671	9675	9680
93	9685	9689	9694	9699	9703	9708	9713	9717	9722	9727
94	9731	9736	9741	9745	9750	9754	9759	9763	9768	9773
95	9777	9782	9786	9791	9795	9800	9805	9809	9814	9818
96	9823	9827	9832	9836	9841	9845	9850	9854	9859	9863
97	9868	9872	9877	9881	9886	9890	9894	9899	9903	9908
98	9912	9917	9921	9926	9930	9934	9939	9943	9948	9952
99	9956	9961	9965	9969	9974	9978	9983	9987	9991	9996
N	0	1	2	3	4	5	6	7	8	9

CLASSIFICATION OF GRAPHS

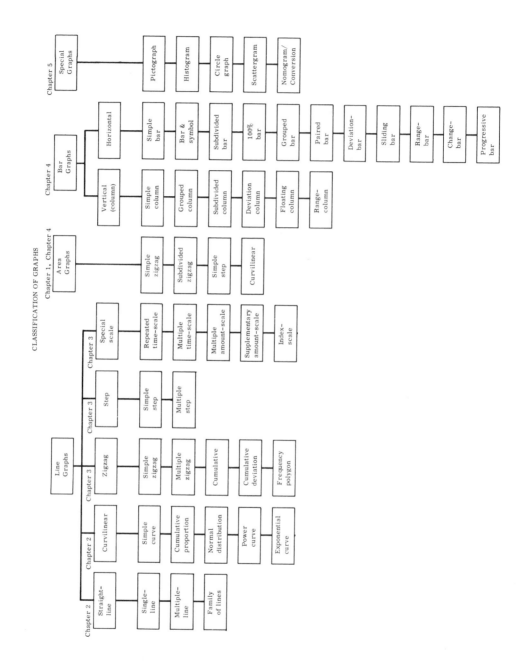

Index